陕西古建筑测绘图辑 泾阳·三原

Shaanxi Historic Buildings Measured and Drawn: Jingyang & Sanyuan

林源 岳岩敏 著

Lin Yuan　Yue Yanmin

中国建筑工业出版社

本书得到"国家艺术基金2017年度传播交流推广资助项目"
[2017-A-03-(143)-0476]的资助。

图书在版编目（CIP）数据

陕西古建筑测绘图辑 泾阳·三原 / 林源, 岳岩敏
著. — 北京：中国建筑工业出版社, 2018.2
ISBN 978-7-112-21722-9

Ⅰ.①陕… Ⅱ.①林…②岳… Ⅲ.①古建筑—建筑测量—泾阳县—图集②古建筑—建筑测量—三原县—图集 Ⅳ.①TU198-64

中国版本图书馆CIP数据核字（2017）第331278号

责任编辑：戚琳琳 率 琦
责任校对：党 蕾

陕西古建筑测绘图辑
泾阳·三原
Shaanxi Historic Buildings Measured and Drawn: Jingyang & Sanyuan
林 源 岳岩敏 著
Lin Yuan Yue Yanmin
*
中国建筑工业出版社出版、发行（北京海淀三里河路9号）
各地新华书店、建筑书店经销
北京京点图文设计有限公司制版
北京中科印刷有限公司印刷
*
开本：787×1092毫米 1/16 印张：14¼ 字数：352千字
2018年3月第一版 2018年3月第一次印刷
定价：49.00元
ISBN 978-7-112-21722-9
（31562）
版权所有 翻印必究
如有印装质量问题，可寄本社退换
（邮政编码 100037）

本书编写人员

编委会：林　源　岳岩敏　喻梦哲　谷瑞超　林　溪　仝梦菲
　　　　雷鸿鹭

撰　　稿：林　源　喻梦哲

英　　文：林　溪　雷鸿鹭　仝梦菲

版式设计：仝梦菲　雷鸿鹭

测　　绘：喻梦哲　林　源　黄思达　申佩玉　夏　楠　岳岩敏
　　　　谷瑞超　孟　玉　李宛儒　姜　川

图纸整理：雷鸿鹭　仝梦菲

Credits

Major Contributors: Lin Yuan, Yue Yanmin, Yu Mengzhe, Gu Ruichao, Lin Xi, Tong Mengfei, Lei Honglu

Texts: Lin Yuan, Yu Mengzhe

English Translation: Lin Xi, Lei Honglu, Tong Mengfei

Book layout: Tong Mengfei, Lei Honglu

Surveyors: Yu Mengzhe, Lin Yuan, Huang Sida, Shen Peiyu, Xia Nan, Yue Yanmin, Gu Ruichao, Meng Yu, Li Wanru, Jiang Chuan

Drawings Preparation: Lei Honglu, Tong Mengfei

序一

建筑遗产对于弘扬民族文化、承续历史文明、增强民族凝聚力和文化认同感具有不可替代的重要作用。陕西在相当长的历史时期内，曾是全国的政治、经济和文化中心，历史积淀极为丰厚，留存至今的建筑遗产种类繁多，数量可观，其中列入全国重点文物保护单位的即近240处，是陕西乃至中华民族悠久历史的珍贵物质见证。

林源教授领衔的建筑历史与理论团队长期致力于对西部广大地域内的建筑遗产进行持续、细致的调查记录与相关研究，在我校几代建筑历史学者辛勤积累的基础上，逐步将西部地区的建筑遗产纳入学界的认知版图。这些工作所取得的最重要成果之一，便是大量的第一手测绘图纸。近年来其团队不断成熟，成果也随之益发丰富，《陕西古建筑测绘图辑》丛书的编纂出版即为对此成果的系统整理。这套丛书可以帮助我们全面、详致地了解具有代表性、典型性的陕西建筑遗产的历史与现状，既是认知、研究及再现历史与文化的可靠学术依据，也是对我校建筑历史与理论学科发展水平的充分展示。

林源教授师从我院著名建筑历史学家赵立瀛教授，现为我院建筑历史与理论学科的学术带头人。近年来其在研究上专注于建筑遗产保护理论研究与实践，在教学上则致力于我院历史建筑保护工程专业的教学体系建设，不仅为学院的学科发展做出了突出贡献，自身也逐渐成长为我国建筑历史与理论学科的中青年领军人物之一。今带领团队著成大作，实属水到渠成之必然，该书亦必将成为陕西地区建筑历史研究的权威著作。

值《陕西古建筑测绘图辑》丛书出版之际，谨表祝贺，是为序。

中国工程院院士　西安建筑科技大学建筑学院院长

丁酉年末于古城西安

Foreword I

by **Prof. Liu Jiaping**
Dean, School of Architecture, XAUAT
Academician, Chinese Academy of Engineering

Architectural heritage has long been playing a non-substitutable role in carrying forward national culture, securing historical continuity, reinforcing ethnical cohesion and enhancing a sense of shared cultural identity. Shaanxi Province was, for a rather long period of time in history, the national political, economic and cultural center. This has bestowed upon us an extraordinarily rich deposit of historic fabrics, which includes an abundance of architectural heritage sites of many types. These sites, of which nearly 240 has been nominated as Key Cultural Heritage Sites under State Protection, now serve as priceless testimonies to the time-honored history of Shaanxi and China.

Led by Prof. Lin Yuan, the Architectural History Section from School of Architecture, Xi'an University of Architecture and Technology (XAUAT), has for decades dedicated themselves in surveying, recording and studying architectural heritage in Western China. Picking up from where the past generations of architectural historians from our university left out, the team has been gradually and steadily incorporating these heritage into the cognitive map of the intellectuals. Among the most valuable results of their endeavors is the multitudinous measured drawings they produced first-hand. The past few years have witnessed the maturing of this team as well as a further accumulation of such yields, which deserve being organized, refined and presented. This is done in the series *Shaanxi Historic Buildings Measured and Drawn*, which offers all-round and in-depth insights into the representative pieces of architectural heritage in Shaanxi, both on their history and current status. The series is expected to, on one hand, provide reliable resource for future investigation, understanding and representation of history and culture to drawn on and on the other, excellently demonstrate the caliber of architectural historians of our university.

Mentored by Prof Zhao Liying, a prominent architectural historian from XAUAT, Prof. Lin Yuan is now the head of our Architectural History Section. For the recent years she has focused on, for research, architectural conservation theories and practice as well as, for teaching, the establishment of the undergraduate program of Historic Building Conservation. Her efforts have not only made distinctive contributions to the school but also accomplished herself as one of the leading figures among Chinese architectural historians of her age. In light of this, the completion of the series is just another success for her and her team from ripe conditions. I firmly believe the series will be an authoritative and indispensible source of reference for researchers on architectural history of Shaanxi.

I hereby applaud the publishing of *Shaanxi Historic Buildings Measured and Drawn* series and congratulate the authors on making it happen.

January, 2018, in Xi'an

(English translation by Lin Xi)

序二

这是西安建筑科技大学建筑历史与理论研究所的老师带领青年学生历经数载，付出辛勤劳动，精心完成的陕西古建筑的调查测绘成果。首次出版的是《陕西古建筑测绘图辑》第一卷——泾阳·三原。

现时代的人们，如何看待古代的建筑？专业工作者、管理部门、社会大众……，难免存在不同的认知和态度。

可喜的是，时至今日，作为主流的、普遍性的认识，都将古代建筑视为民族的物质文化遗产，视为宝贵的物质文化资源，不仅具有历史价值，而且具有现代价值，应当给予科学的保护与合理的利用。很少有人将古代建筑遗存视为城乡新建设的"包袱"、"障碍"，主张拆除、平毁。

古代建筑的遗产，诸如城市中的历史街区、古建筑、传统民居；古镇、古村落；风景名胜区中的古建筑等等，如何在城市更新中、乡村改造中、风景名胜区建设中以及现代建筑创作中得到包容、传承，焕发出新的生机，体现出新的价值，是摆在人们面前的研究与实践的双重课题。

我们需要认真的、踏实的工作。许多事情，只有理解得越深，方能做得越好。如此，才有助于真正做好古建筑遗产的科学保护与合理利用。

古建筑，首先具有历史性，产生于特定的历史年代，又具有地域性，产生于特定的地域环境；涵盖技术（材料、构造、结构、工艺……），科学（环境、地形、气候、水流、风向、防灾……），文化（哲学、美学、习俗、信仰、人文价值观与审美观……），艺术（空间、造型、风格、装饰……）等方方面面的丰富内容。

本书作者所做的调查测绘工作和成果，是古建筑研究的最为必要的基础性工作。缺少这个基础，一切"高谈阔论"，可以说，都将是无根之木、无源之水！《陕西古建筑测绘图辑》的陆续完成、出版，应当说，将有助于陕西古建筑研究的深化与提高。

在我国，面对如此广大的地域，大量的古代建筑还有历史建筑遗存，古建筑的研究工作和保护工程技术人才培养的状况，都还远远跟不上当前与未来的需求。

在《陕西古建筑测绘图辑》首卷出版之际，写了上面这些话，以为序。

西安建筑科技大学建筑学院教授　博士生导师

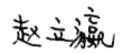

于 2018 年 1 月 20 日

Foreword II

by **Prof. Zhao Liying**
Ph.D. Candidate Supervisor, XAUAT, retired.
Founder of the Architectural History Section, XAUAT

What lies now before you is years of hard work having come to fruition. The faculty and students from Architectural History Section, Xi'an University of Architecture and Technology (XAUAT) undertook this work with assiduous labor and painstaking attention. Thus they are finally able to present to you the results from surveying and measured drawing on ancient architectural works in Shaanxi Province in this Volume, *Jinyang & Sanyuan* the first to have been published among the series *Shaanxi Historic Buildings Measured and Drawn*.

What are the architectural works from our past to us today? People from various fields, say professionals, administrators and the general public, etc., are bound to have different answers to this question.

However, the good news is that it has become a mainstream and universally-accepted notion to recognize ancient architectural works as invaluable material as well as cultural legacy and resource of our nation. These works are of value to both the past and the present, hence deserving scientific preservation and proper utilization. Few people would argue that these remains of the past are liabilities or obstacles to our rural-urban development and therefore should be razed to the ground.

Ancient architectural heritage may include historic buildings, vernacular dwellings and areas in urban settings, as well as historic towns, villages, plus architectural works in scenic areas. To incorporate, inherit, re-enliven and re-valorize them in urban renewal, rural reformation, scenic area development and contemporary architectural designing is nowadays posing a challenge faced by both researchers and practitioners.

This calls for steadfastness and seriousness in our work. In many senses, only deeper understandings could breed better judgements. Scientific preservation and proper utilization of these legacies could be achieved by no other way.

Ancient architectural works are first spatio-temporally bound, that is, situated in a historical period and a locality which are both specific. Second, these works were informed by and imbued with a multi-faceted richness of inputs and messages, be they, including but not limited to, technical (materials, constructions, structures, craftsmanship, etc.), scientific (environments, topographies, climates, natural elements, risk preparedness, etc.), cultural (ideologies, aesthetics, customs, beliefs, value systems, etc.), and artistic (space, form, style, ornament etc.).

The surveying and measured drawing tasks completed by the authors of this book is the most essential groundwork for studying ancient architecture. Any discourse and declamation on the subject in the absence of such kind of work will be an attempt to make omelets without eggs. Eventual completion and publication of the volumes in the series will, therefore, be conducive to the furthering and enhancing of studies on ancient architecture in Shaanxi.

Considering that China has so many sites of architectural heritage, both ruinous and standing, scattered over such a vast territory, I regret the fact that our work in the field, including research, practice and personnel training, still has a very long way to go before the present and future demands could be met.

With the publication of the first volume in the series, I hereby present these words and pray they be of some benefit to those who read it.

January 20th, 2018

(English translation by Lin Xi)

目　录

序一
序二

图纸		001	
泾阳			
01	[唐] 唐德宗崇陵石刻	002	(176)
02	[唐] 唐宣宗贞陵石刻	012	(176)
03	[明-清] 泾阳文庙	020	(177)
04	[明] 太壸寺大雄殿	052	(177)
05	[清] 迎祥宫戏台	064	(178)
06	[清] 味经书院	074	(179)
07	[清] 崇实书院	076	(179)
08	[中华民国] 韩家堡小学	078	(180)
09	[唐] 振锡寺悟空禅师塔	080	(180)
10	[明] 崇文塔	084	(180)
11	[清] 张家屯村大义坊	090	(181)
12	[清] 李家村防卫楼	094	(181)
13	[清] 李仪祉故居	100	(182)
14	[中华民国] 赵春喜民居	110	(182)
15	[清] 蒋明杰民居	112	(182)
16	[中华民国] 高兰亭故居	116	(183)
17	[清] 吴氏庄园	118	(183)
18	[中华民国] 望月楼	126	(184)
三原			
19	[清] 孟店周宅	128	(184)
20	[明] 三原城隍庙	140	(184)

说明文字	175
参考文献	194
附录	195
附录一　本书收录的陕西泾阳、三原地区现存古建筑一览表	196
附录二　图纸目录	198
后　记	210

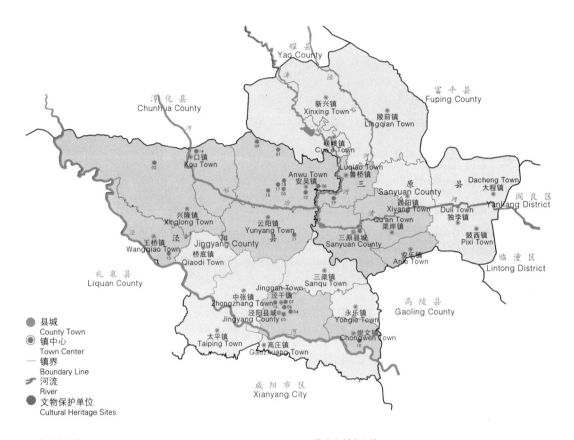

01 唐德宗崇陵 Stone Sculptures, Chong Mausoleum of Tang	11 张家屯村大义坊 Dayi Pailou, Zhangjiatun Village
02 唐宣宗贞陵 Stone Sculptures, Zhen Mausoleum of Tang	12 李家村防卫楼 Defensive Building, Lijia Village
03 泾阳文庙 Confucian Temple of Jingyang	13 李仪祉故居（中华水利会馆） Former Residence of Li Yizhi
04 太壶寺大雄殿 Mahavira Hall of Taikun Temple	14 赵春喜民居 Residence of Zhao Chunxi
05 迎祥宫戏台 Opera stage of Yingxiang Palace	15 蒋明杰民居 Former Residence of Jiang Mingjie
06 味经书院 Weijing Academy	16 高兰亭故居 Former Residence of Gao Lanting
07 崇实书院 Chongshi Academy	17 吴氏庄园 Wu Manor
08 韩家堡小学 Primary School of Hanjiapu	18 望月楼 Wangyue Building
09 振锡寺悟空禅师塔 Zen-Master Wukong Pagoda	19 孟店周宅 Zhou Residence of Mengdian
10 崇文塔 Chongwen Pagoda	20 三原城隍庙 Cheng-Huang Temple of Sanyuan

x

Contents

Foreword I

Foreword II

Drawings	001
Jingyang	
01 [Tang] Stone Sculptures, Chong Mausoleum of Tang	002 (185)
02 [Tang] Stone Sculptures, Zhen Mausoleum of Tang	012 (185)
03 [Ming-Qing] Confucian Temple of Jingyang	020 (186)
04 [Ming] Mahavira Hall of Taikun Temple	052 (186)
05 [Qing] Opera Stage of Yingxiang Palace	064 (187)
06 [Qing] Weijing Academy	074 (187)
07 [Qing] Chongshi Academy	076 (187)
08 [ROC] Primary School of Hanjiapu	078 (188)
09 [Tang] Zen-Master Wukong Pagoda	080 (188)
10 [Ming] Chongwen Pagoda	084 (189)
11 [Qing] Dayi *Pailou*, Zhangjiatun Village	090 (189)
12 [Qing] Defensive Building, Lijia Village	094 (189)
13 [Qing] Former Residence of Li Yizhi	100 (190)
14 [ROC] Residence of Zhao Chunxi	110 (190)
15 [Qing] Former Residence of Jiang Mingjie	112 (190)
16 [ROC] Former Residence of Gao Lanting	116 (191)
17 [Qing] Wu Manor	118 (191)
18 [ROC] Wangyue Building	126 (192)
Sanyuan	
19 [Qing] Zhou Residence of Mengdian	128 (192)
20 [Ming] Cheng-Huang Temple of Sanyuan	140 (192)
About the Sites	175
References	194
Appendixes	195
Appendix I : List of Sites	197
Appendix II : List of Drawings	204
Postscript	212

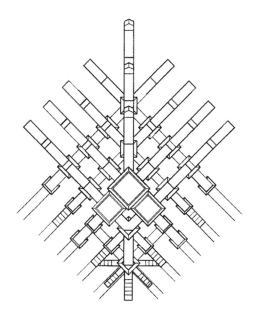

图　纸

Drawings

陕西古建筑测绘图辑
泾阳·三原

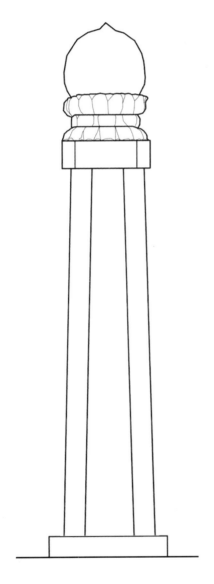

| 入口 Entrance | | 通向陵体 Leading to the tomb |

01 [唐]唐德宗崇陵石刻
01 [Tang] Stone Sculptures, Chong Mausoleum of Tang

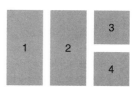

1. 华表立面图
   Elevation, a *huabiao*
2. 华表现状照片
   A *huabiao* (photo)
3. 翼马现状照片
   A winged horse (photo)
4. 文臣现状照片
   A civil servant (photo)

陕西古建筑测绘图辑
泾阳・三原

入口
Entrance

通向陵体
Leading to the tomb

1. 翼马正面
   Front elevation, a winged horse
2. 翼马侧面
   Side elevation, a winged horse

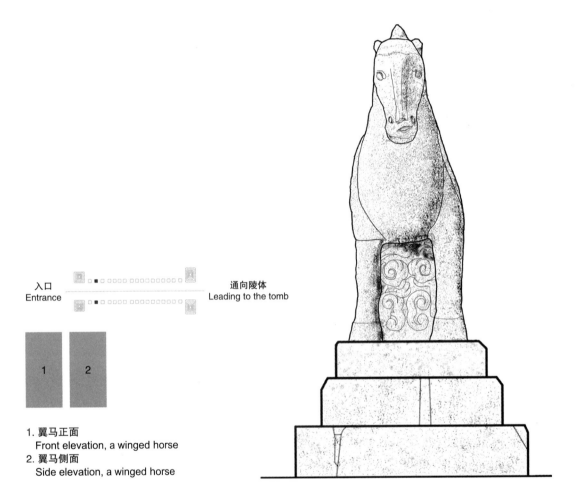

01　[唐]唐德宗崇陵石刻
01　[Tang] Stone Sculptures, Chong Mausoleum of Tang

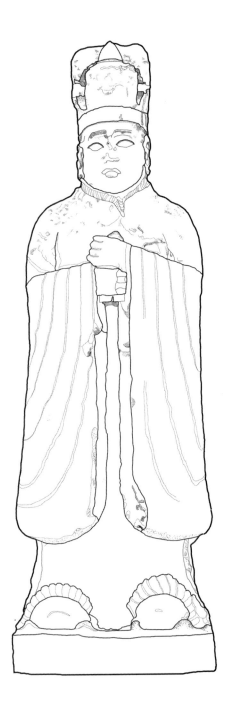

01 ［唐］唐德宗崇陵石刻

01　[Tang] Stone Sculptures, Chong Mausoleum of Tang

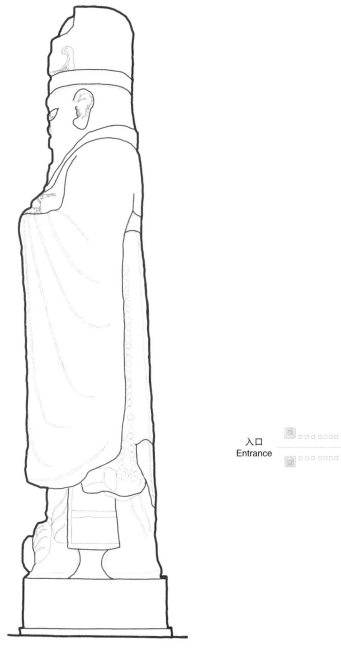

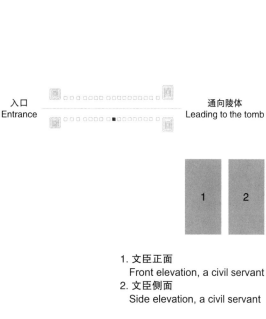

入口
Entrance

通向陵体
Leading to the tomb

1. 文臣正面
   Front elevation, a civil servant
2. 文臣侧面
   Side elevation, a civil servant

陕西古建筑测绘图辑

泾阳·三原

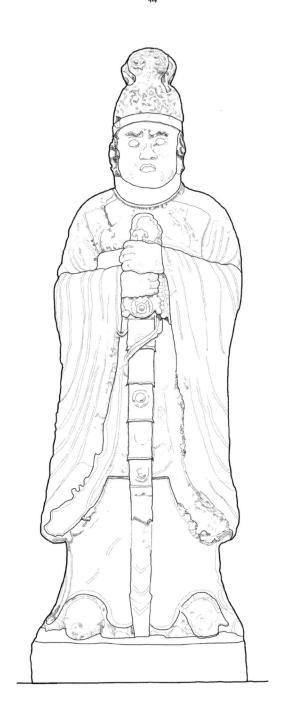

01 [唐]唐德宗崇陵石刻
01 [Tang] Stone Sculptures, Chong Mausoleum of Tang

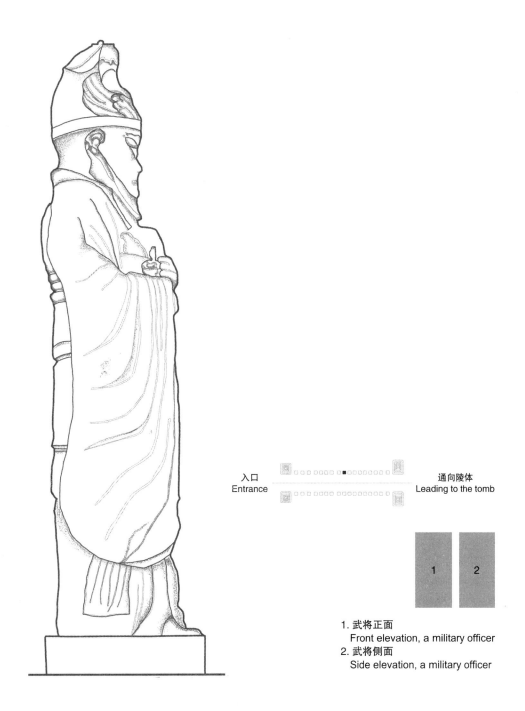

入口
Entrance

通向陵体
Leading to the tomb

1. 武将正面
   Front elevation, a military officer
2. 武将侧面
   Side elevation, a military officer

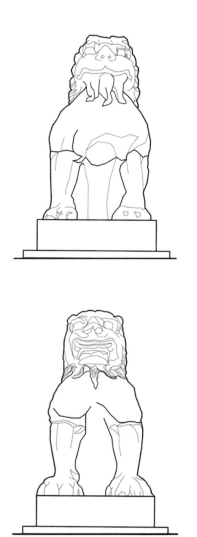

01 [唐]唐德宗崇陵石刻

01 [Tang] Stone Sculptures, Chong Mausoleum of Tang

入口　　　　　　　　　　　　通向陵体
Entrance　　　　　　　　　　Leading to the tomb

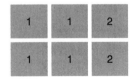

1. 石狮大样图
   Details, a lion
2. 石狮现状照片
   A lion (photo)

陕西古建筑测绘图辑
泾阳・三原

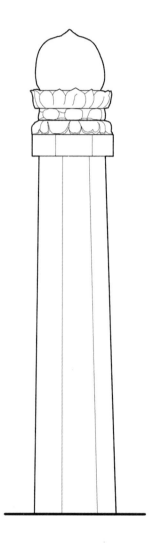

| 入口 | | 通向陵体 |
| Entrance | | Leading to the tomb |

02 [唐]唐宣宗贞陵石刻
02 [Tang] Stone Sculptures, Zhen Mausoleum of Tang

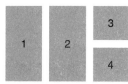

1. 华表立面图
   Elavation, a *huabiao*
2. 华表现状照片
   A *huabiao* (photo)
3. 翼马现状照片
   A winged horse (photo)
4. 石人现状照片
   Stone figures (photo)

陕西古建筑测绘图辑
泾阳·三原

02 [唐]唐宣宗贞陵石刻
02 [Tang] Stone Sculptures, Zhen Mausoleum of Tang

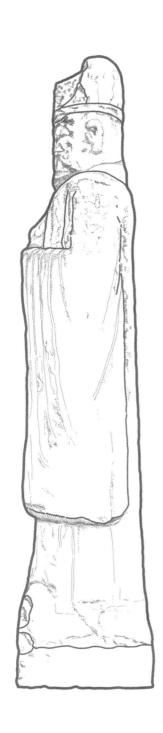

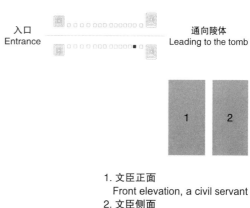

入口　Entrance　　通向陵体　Leading to the tomb

1. 文臣正面
   Front elevation, a civil servant
2. 文臣侧面
   Side elevation, a civil servant

陕西古建筑测绘图辑
泾阳·三原

02　[唐] 唐宣宗贞陵石刻
02　[Tang] Stone Sculptures, Zhen Mausoleum of Tang

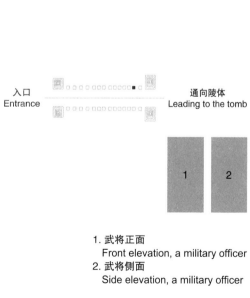

入口　Entrance　　通向陵体　Leading to the tomb

1. 武将正面
   Front elevation, a military officer
2. 武将侧面
   Side elevation, a military officer

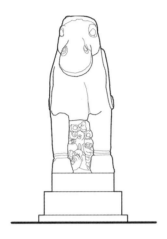
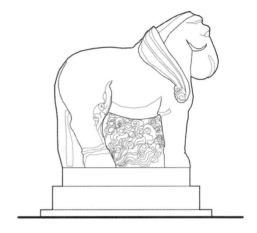

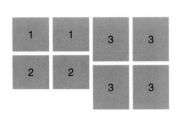

1. 翼马大样图
   Details, a winged horse
2. 卧马大样图
   Details, a crouching horse
3. 石刻现状照片
   Stone sculptures (photos)

入口　Entrance　　　通向陵体　Leading to the tomb

02 [唐]唐宣宗贞陵石刻
02 [Tang] Stone Sculptures, Zhen Mausoleum of Tang

陕西古建筑测绘图辑
泾阳·三原

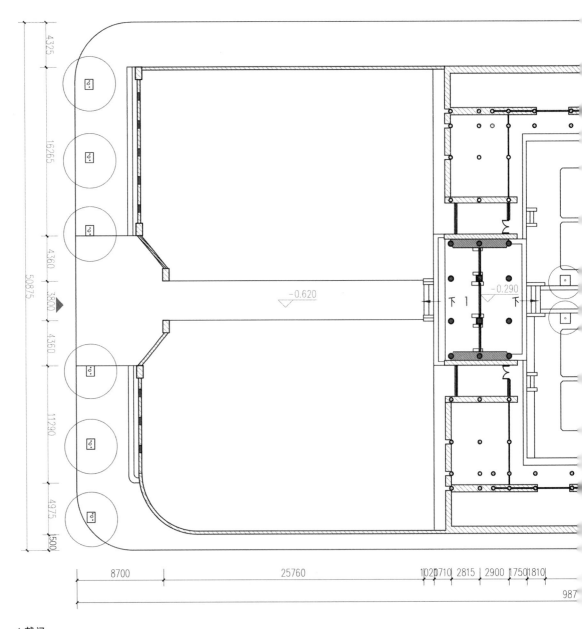

1. 戟门
   Ji-men
2. 大成殿
   Dacheng-Dian

03 [明-清]泾阳文庙
03 [Ming–Qing] Confucian Temple of Jingyang

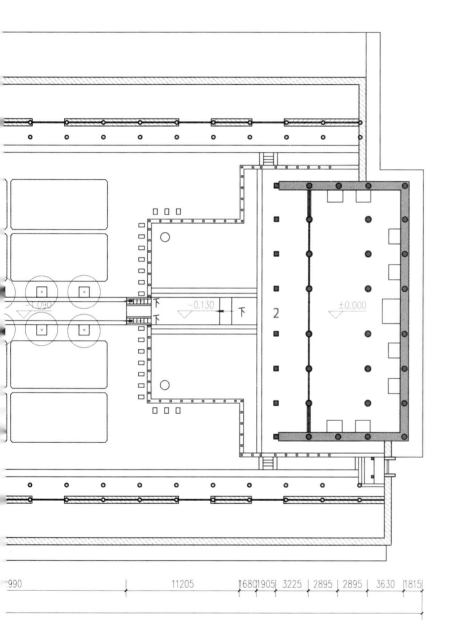

1. 总平面图
Site plan

陕西古建筑测绘图辑
泾阳·三原

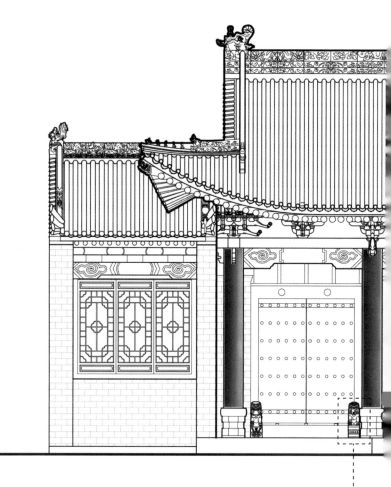

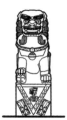 

1. 戟门南立面图
   South elevation, Ji-men
2. 戟门石狮大样图
   Details, a stone lion at Ji-men
3. 戟门抱鼓石大样图
   Details, a drum-stone of Ji-men

03　[明-清]泾阳文庙

03　[Ming-Qing] Confucian Temple of Jingyang

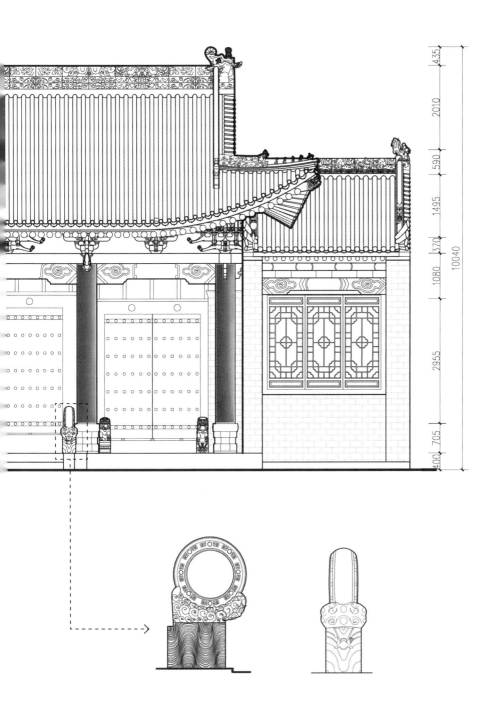

陕西古建筑测绘图辑

泾阳·三原

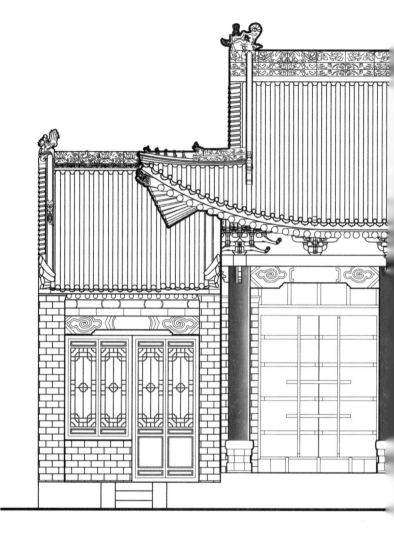

1. 戟门北立面图
   North elevation, *Ji-men*
2. 戟门瓦当大样图
   Details, eaves tiles of *Ji-men*
3. 戟门滴水大样图
   Details, a flashing tile of *Ji-men*

03 [明–清]泾阳文庙
03 [Ming–Qing] Confucian Temple of Jingyang

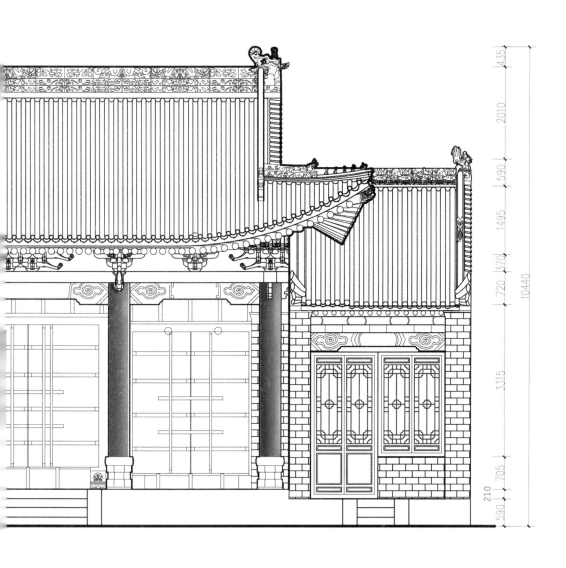

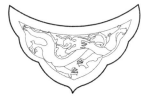

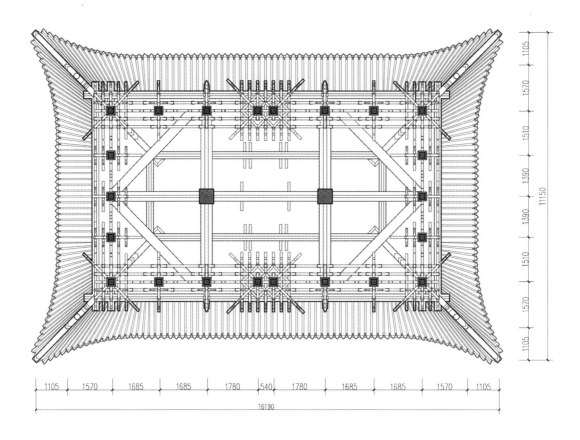

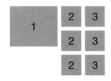

1. 戟门梁架仰视图
   Reflected plan, the frame structure of *Ji-men*
2. 戟门当心间平身科大样图
   Details, an intermediate *dougong* set at the central bay, *Ji-men*
3. 戟门次间平身科斗拱大样图
   Details, an intermediate *dougong* set at a second-to-the-central bay, *Ji-men*

03 [明-清]泾阳文庙

03 [Ming-Qing] Confucian Temple of Jingyang

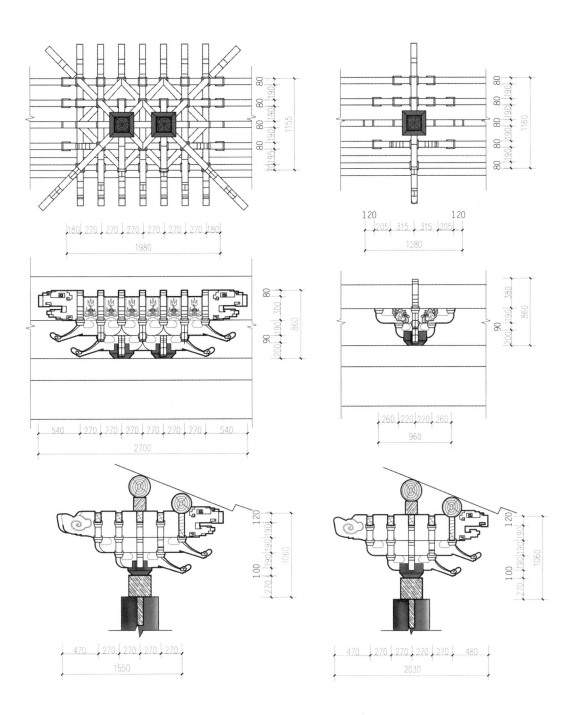

陕西古建筑测绘图辑

泾阳·三原

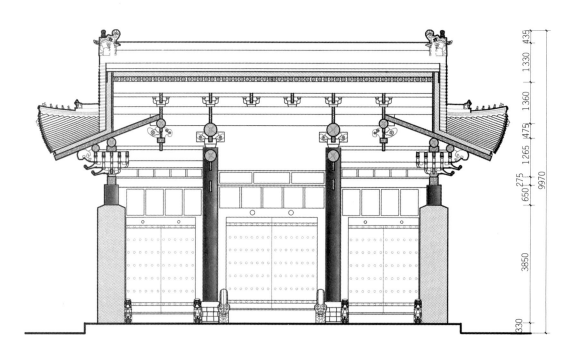

03　[明–清]泾阳文庙
03　[Ming–Qing] Confucian Temple of Jingyang

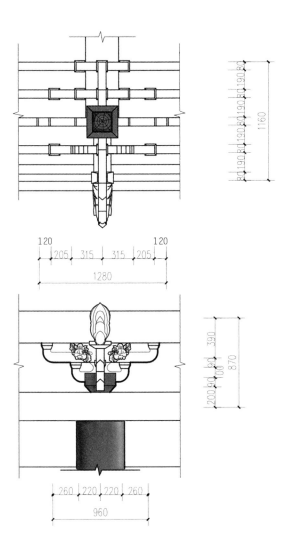

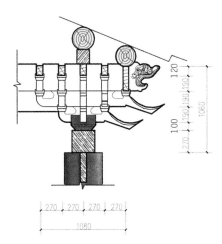

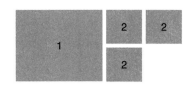

1. 戟门纵剖面图
   Longitudinal section, *Ji-men*
2. 戟门柱头科大样图
   Detail, an on-column *dougong* set, *Ji-men*

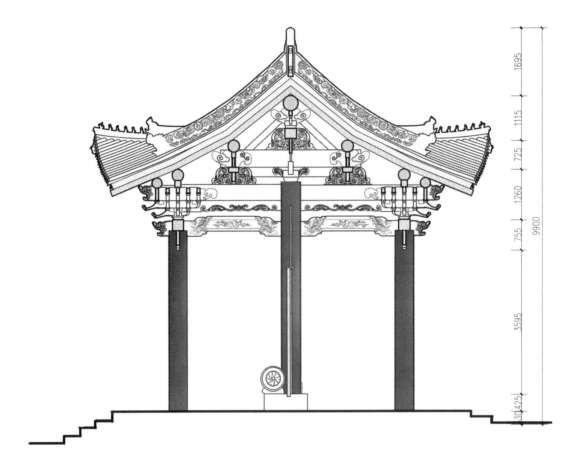

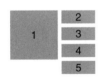

1. 戟门横剖面图
   Transverse section, Ji-men
2. 戟门脊檩下驼峰大样图
   Details, a *tuofeng* (camel-hump shaped support) beneath the ridge purlin, Ji-men
3. 戟门金檩下驼峰大样图
   Details, a *tuofeng* beneath an intermediate purlin, Ji-men
4. 戟门随梁枋大样图
   Details, a tie-beam along the lower edge of a crescent beam, Ji-men
5. 戟门额枋大样图
   Details, an architrave, Ji-men

03 [明-清]泾阳文庙
03 [Ming-Qing] Confucian Temple of Jingyang

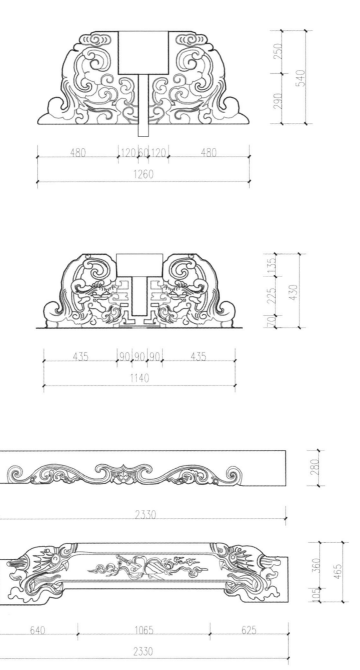

陕西古建筑测绘图辑
泾阳·三原

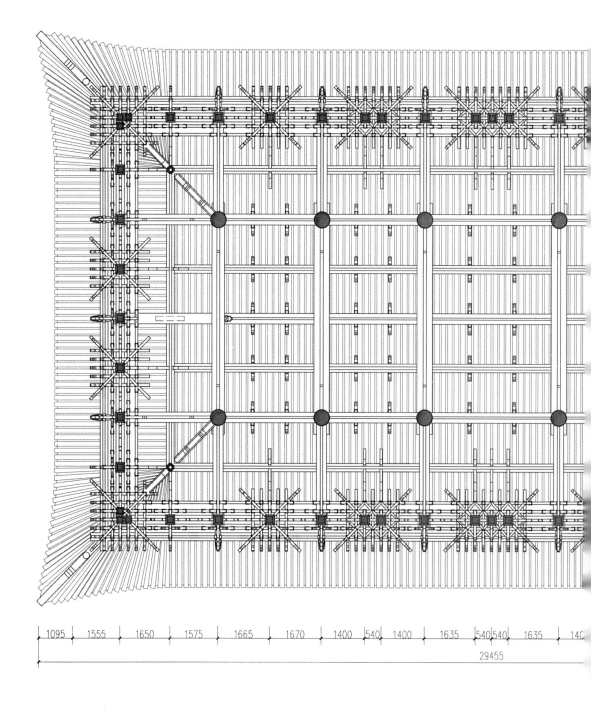

03 [明-清]泾阳文庙
03 [Ming-Qing] Confucian Temple of Jingyang

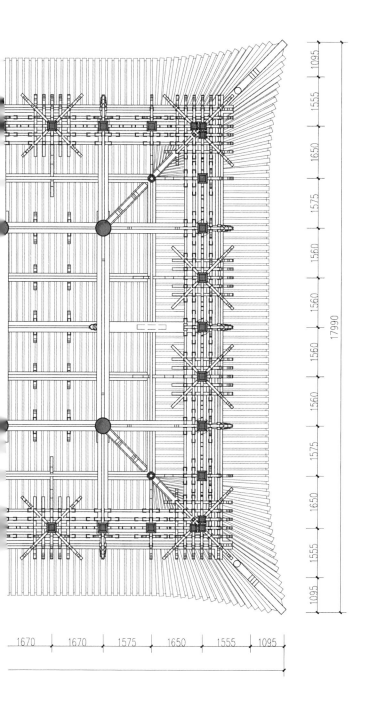

1. 大成殿梁架仰视图
Reflected plan, the frame structure of *Dacheng-Dian*

陕西古建筑测绘图辑

泾阳 · 三原

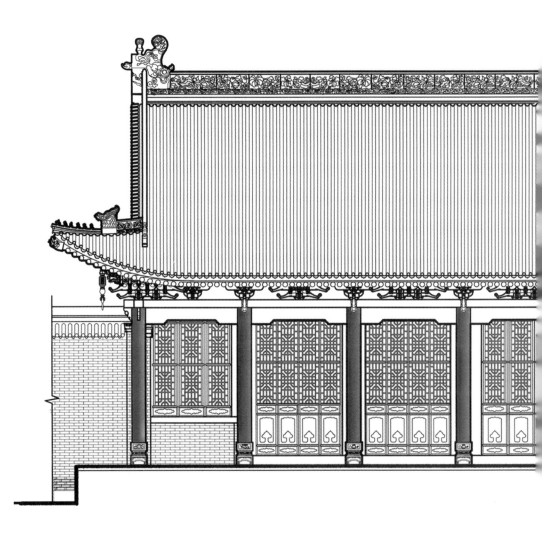

03 [明–清]泾阳文庙
03 [Ming–Qing] Confucian Temple of Jingyang

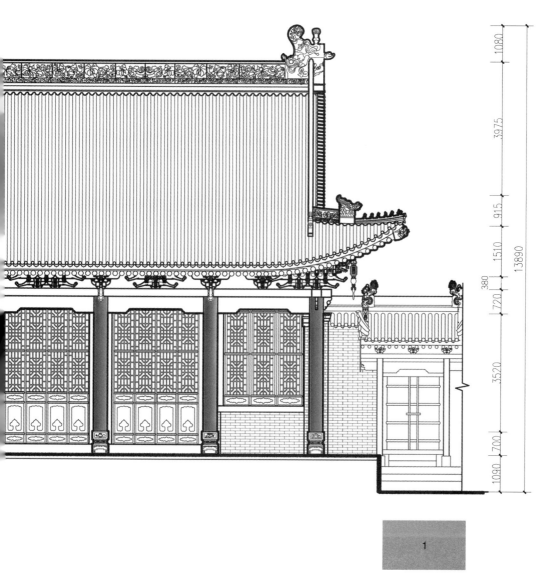

1

1. 大成殿南立面图
South elevation, *Dacheng-Dian*

陕西古建筑测绘图辑

泾阳·三原

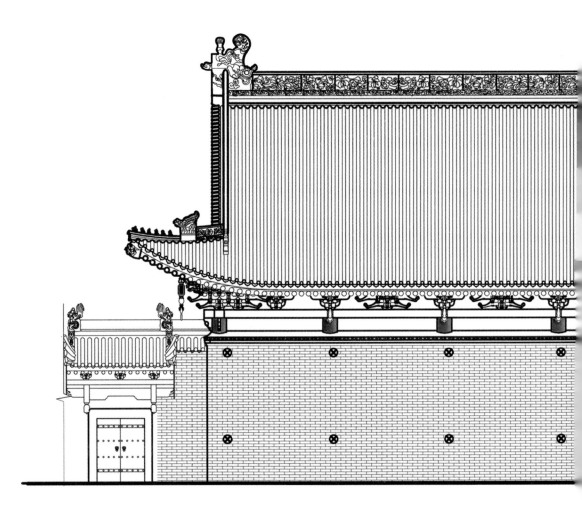

03 [明－清]泾阳文庙
03 [Ming–Qing] Confucian Temple of Jingyang

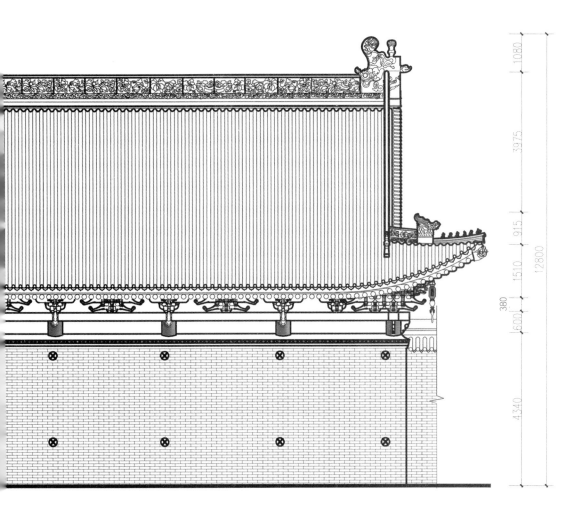

1. 大成殿北立面图
   North elevation, *Dacheng-Dian*

陕西古建筑测绘图辑

泾阳·三原

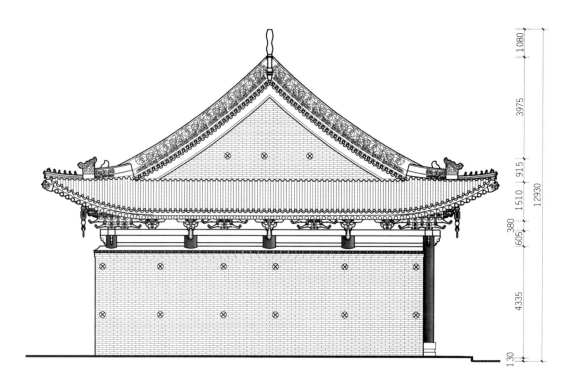

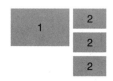

1. 大成殿西立面图
   West elevation, *Dacheng-Dian*
2. 大成殿次间柱头科大样图
   Details, an on-column *dougong* set at a secondary bay, *Dacheng-Dian*

03 [明-清]泾阳文庙
03 [Ming–Qing] Confucian Temple of Jingyang

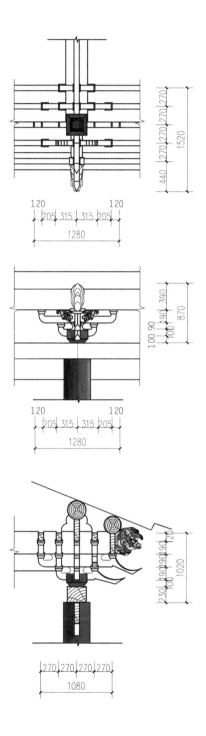

陕西古建筑测绘图辑
泾阳·三原

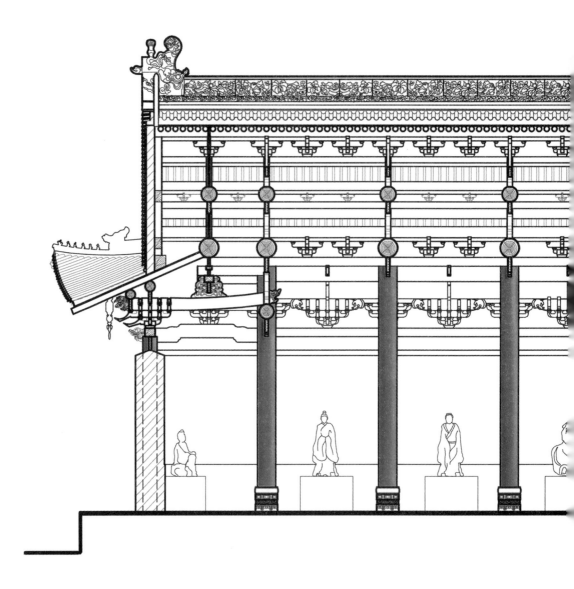

03　[明－清] 泾阳文庙

03　[Ming-Qing] Confucian Temple of Jingyang

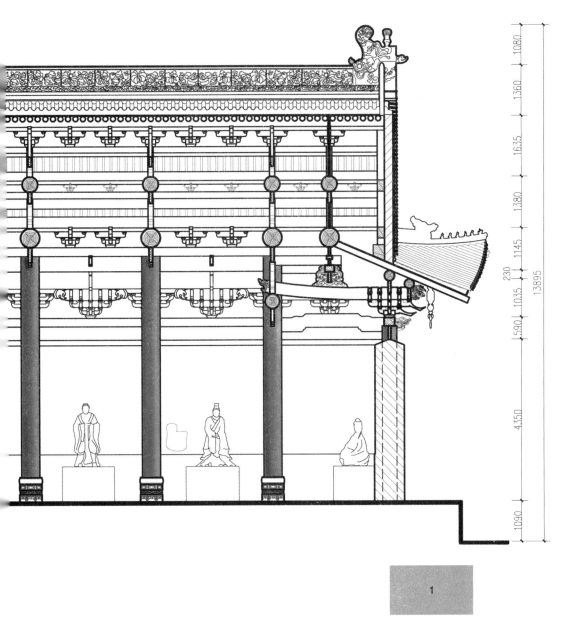

1. 大成殿纵剖面图
   Longitudinal section, *Dacheng-Dian*

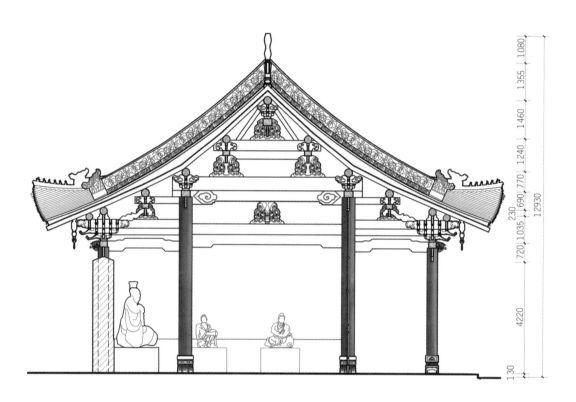

03 [明-清] 泾阳文庙
03 [Ming-Qing] Confucian Temple of Jingyang

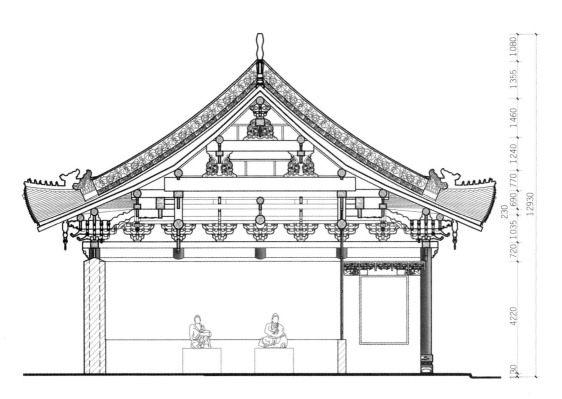

1. 大成殿当心间横剖面图
   Transverse section, at the central bay, *Dacheng-Dian*
2. 大成殿梢间横剖面图
   Transverse section, at a lateral bay, *Dacheng-Dian*

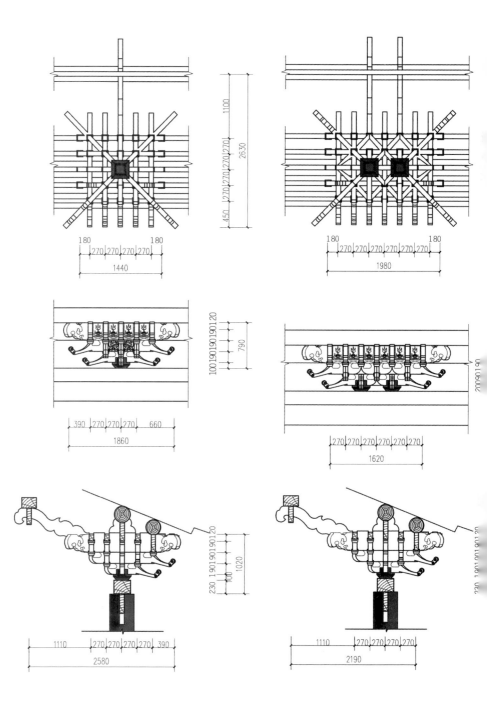

03 [明-清]泾阳文庙
03 [Ming-Qing] Confucian Temple of Jingyang

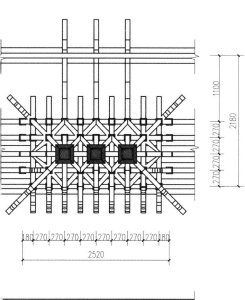

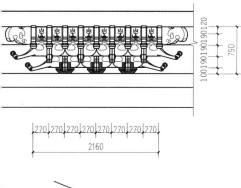

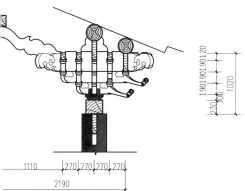

1. 大成殿当心间平身科大样图
   Details, an intermediate *dougong* set at the central bay, Dacheng-Dian
2. 大成殿次间平身科大样图
   Details, an intermediate *dougong* set at a third-to-the-central bay, Dacheng-Dian
3. 大成殿梢间平身科大样图
   Details, an intermediate *dougong* set at a lateral bay, Dacheng-Dian

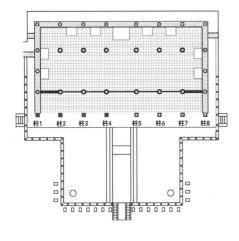
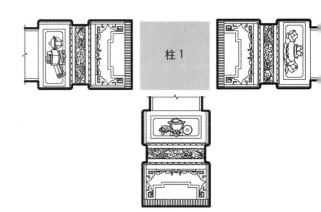
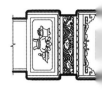

1. 大成殿柱础大样图
   Details, column bases, *Dacheng-Dian* (Part 1)

03 [明-清]泾阳文庙
03 [Ming–Qing] Confucian Temple of Jingyang

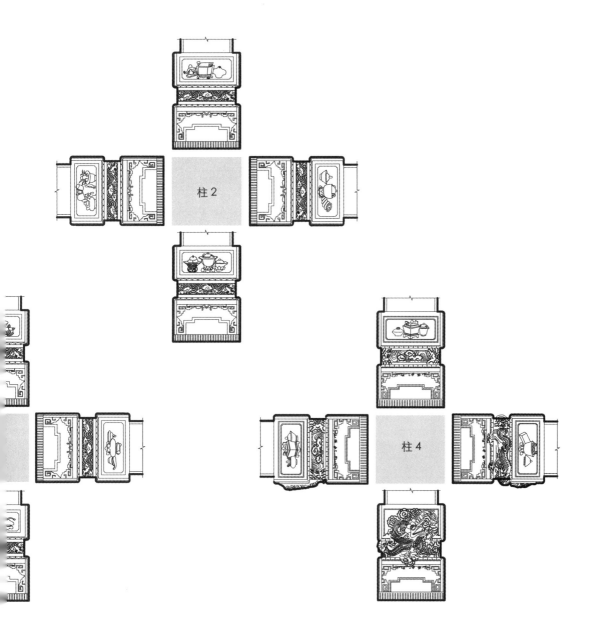

陕西古建筑测绘图辑
泾阳·三原

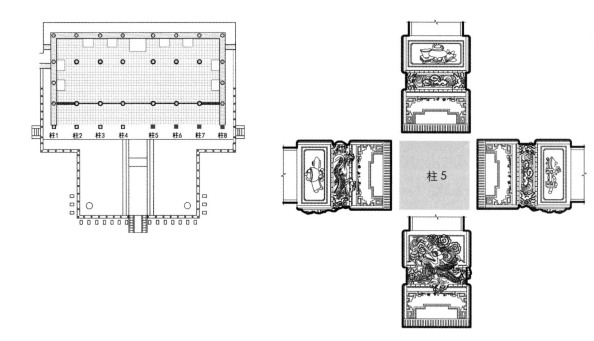

1. 大成殿柱础大样图
   Details, column bases, *Dacheng-Dian* (Part 2)

03 ［明－清］泾阳文庙
03 [Ming-Qing] Confucian Temple of Jingyang

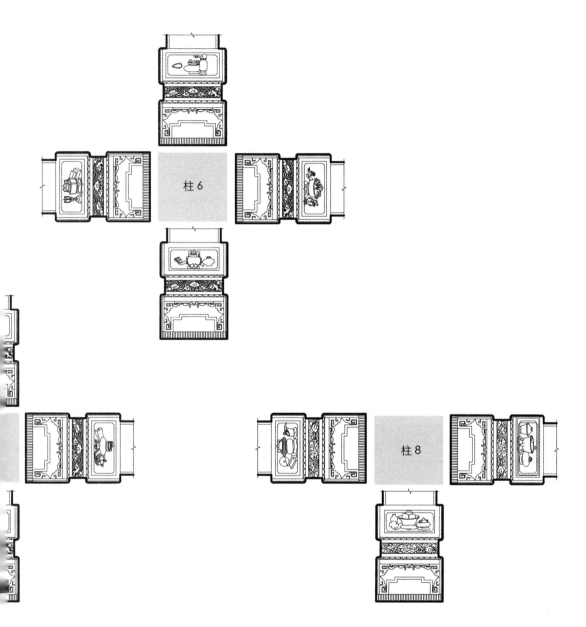

戟门正面
Front, *Ji-men* (photo)

戟门背面
Back, *Ji-men* (photo)

大成殿正面
Front, *Dacheng-Dian* (photo)

大成殿翼角
An eave corner, *Dacheng-Dian* (photo)

03 ［明-清］泾阳文庙
03 [Ming–Qing] Confucian Temple of Jingyang

大成殿前檐当心间前檐
Entablature, the central bay, *Dacheng-Dian* (photo)

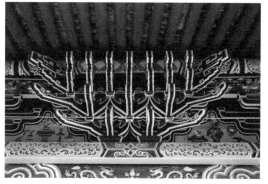

大成殿前檐补间斗栱
An inter-column *dougong* set, *Dacheng-Dian* (photo)

陕西古建筑测绘图辑

泾阳·三原

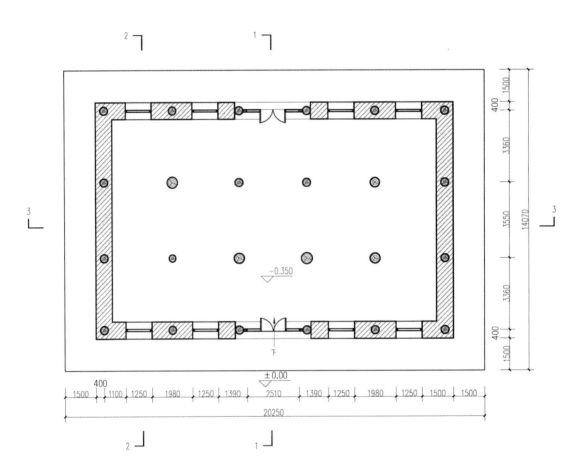

04 [明] 太壶寺大雄殿
04 [Ming] Mahavira Hall of Taikun Temple

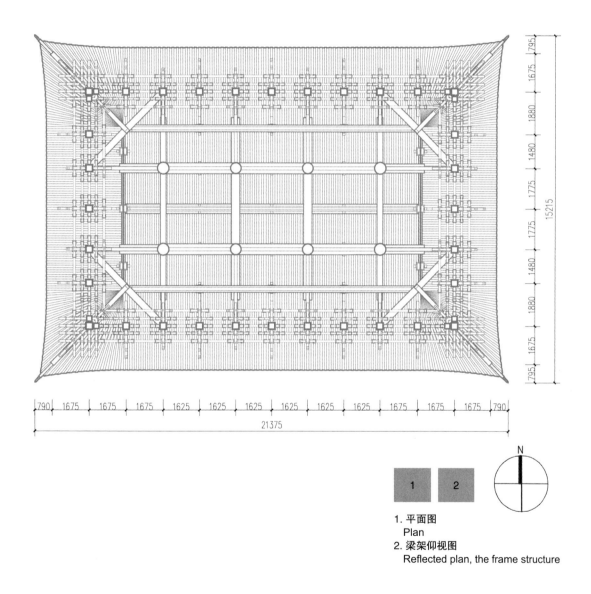

1. 平面图
   Plan
2. 梁架仰视图
   Reflected plan, the frame structure

陕西古建筑测绘图辑

泾阳·三原

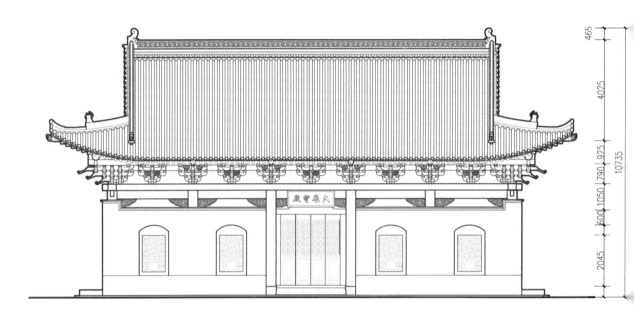

04 [明]太壶寺大雄殿
04 [Ming] Mahavira Hall of Taikun Temple

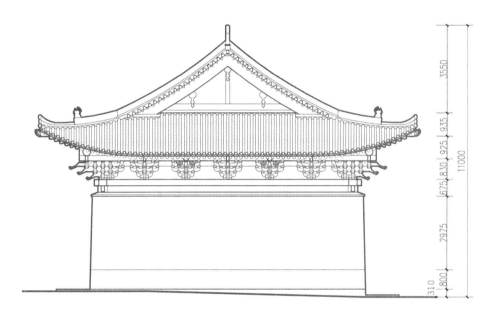

1. 南立面图
   South elevation
2. 东立面图
   East elevation

陕西古建筑测绘图辑
泾阳·三原

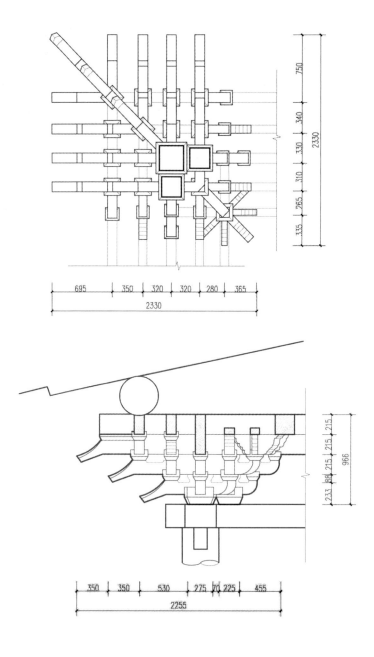

04 [明] 太壶寺大雄殿
04 [Ming] Mahavira Hall of Taikun Temple

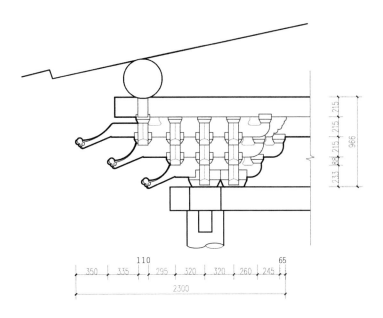

1. 转角铺作大样图
   Details, an on-corner *dougong* set

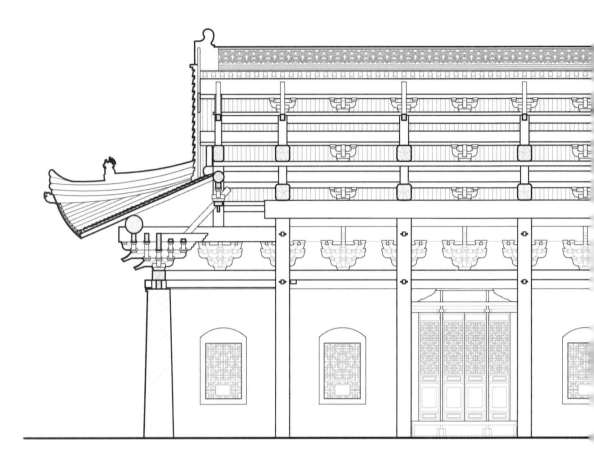

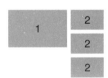

1. 3–3 剖面图
   Section 3-3
2. 柱头铺作大样图
   Details, an on-column *dougong* set

04 [明] 太壺寺大雄殿
04 [Ming] Mahavira Hall of Taikun Temple

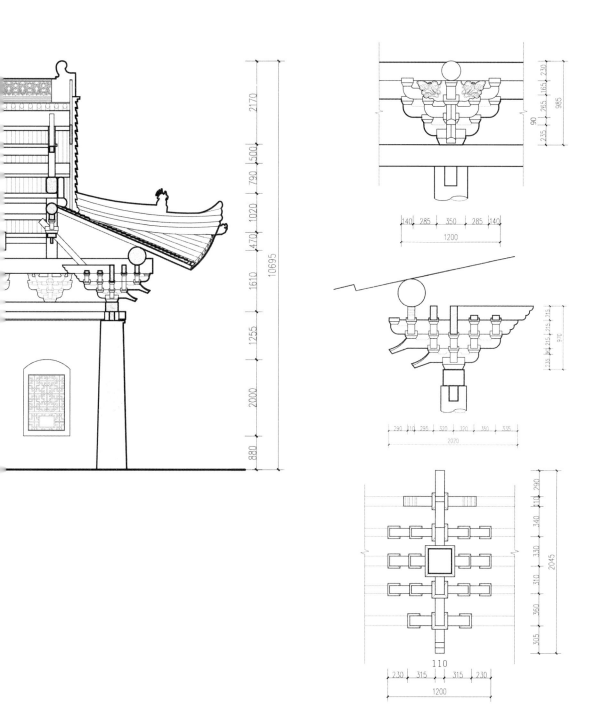

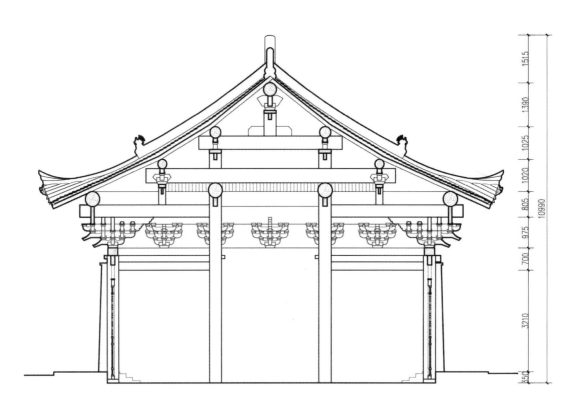

1. 1–1 剖面图
   Section 1-1
2. 2–2 剖面图
   Section 2-2

04 [明]太壶寺大雄殿
04 [Ming] Mahavira Hall of Taikun Temple

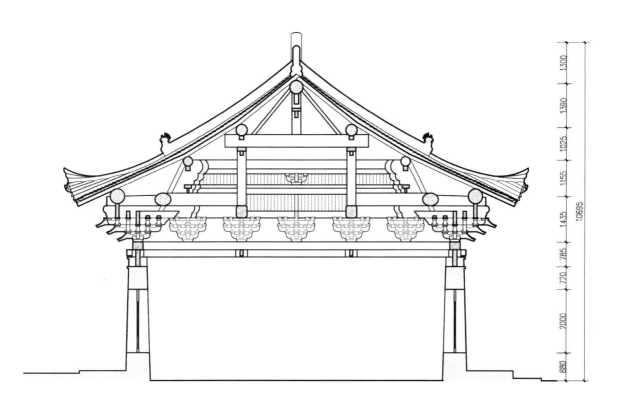

陕西古建筑测绘图辑
泾阳·三原

04 [明]太壶寺大雄殿
04 [Ming] Mahavira Hall of Taikun Temple

陕西古建筑测绘图辑

泾阳·三原

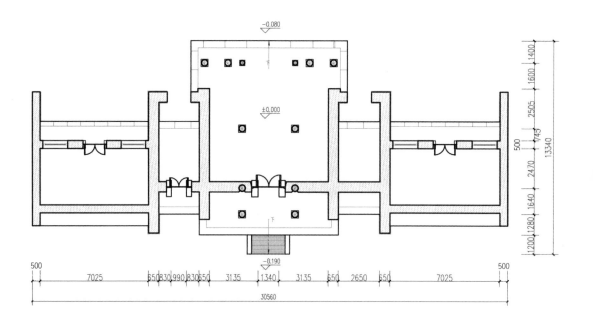

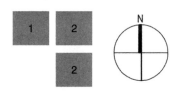

1. 一层平面图
   1F plan
2. 北立面当心间平身科大样图
   An intermediate *dougong* set at the central bay, north façade

05 [清]迎祥宫戏台

05 [Qing] Opera Stage of Yingxiang Palace

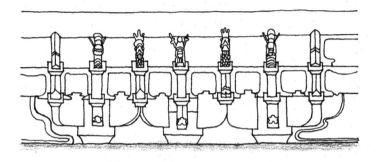

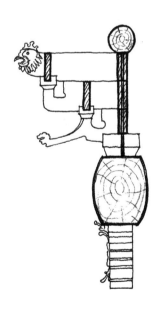

陕西古建筑测绘图辑
泾阳·三原

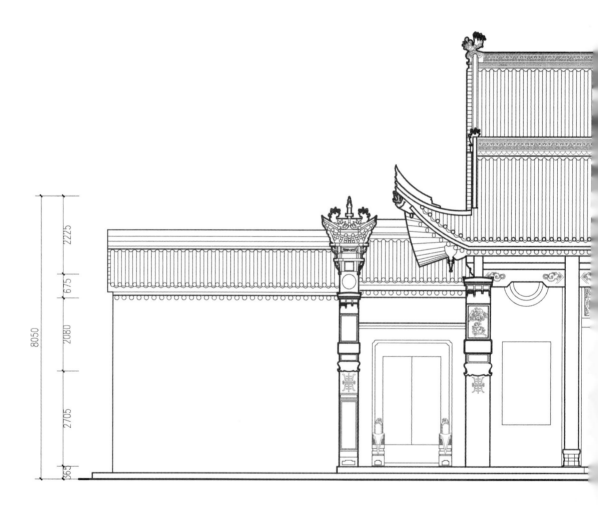

05　[清]迎祥宫戏台
05　[Qing] Opera Stage of Yingxiang Palace

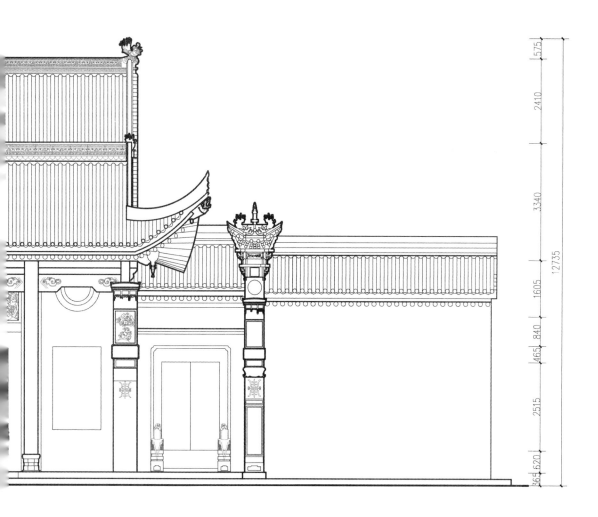

1. 南立面图
South elevation

陕西古建筑测绘图辑

泾阳·三原

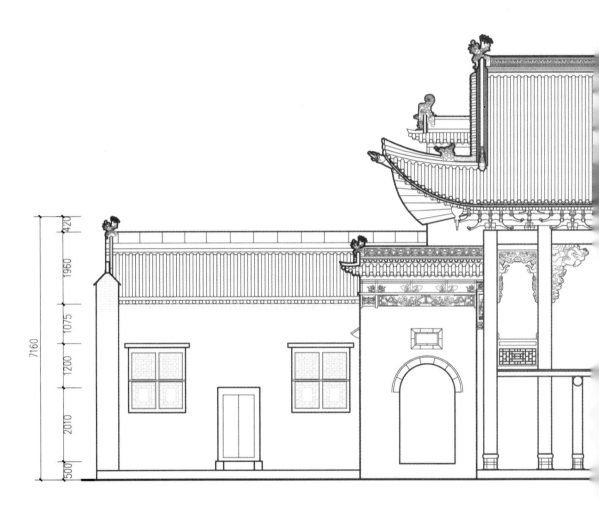

05　[清]迎祥宫戏台

05　[Qing] Opera Stage of Yingxiang Palace

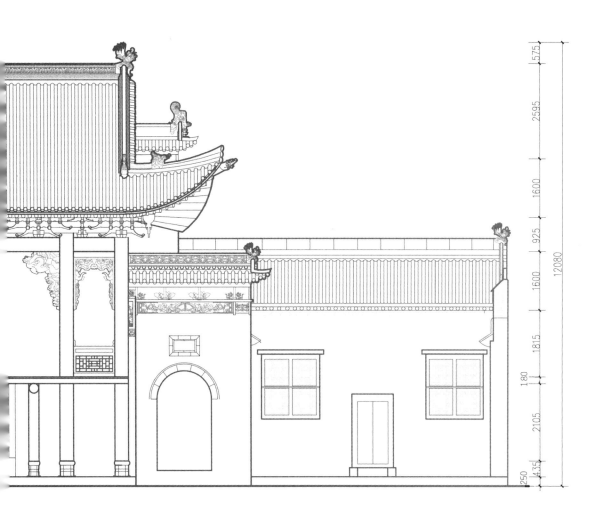

1. 北立面图
North elevation.

陕西古建筑测绘图辑

泾阳·三原

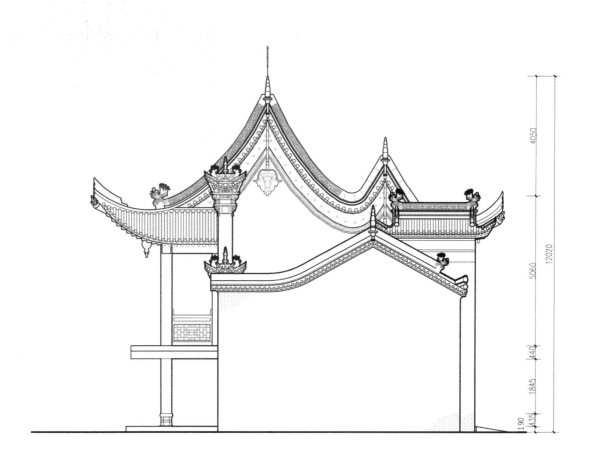

05 [清]迎祥宫戏台
05 [Qing] Opera Stage of Yingxiang Palace

1. 西立面图
   West elevation
2. 砖雕墙头大样图
   Details, the carved-brick wall header.
3. 吻兽大样图
   Details, a *wenshou* (on-roof beast-shaped ornament).

入口
Entrance (photo)

入口匾额
Name plague at the entrance (photo)

入口石狮
A stone lion at the entrance (photo)

入口门砧
A bearing-stone for the entrance (photo)

05 [清]迎祥宫戏台
05 [Qing] Opera Stage of Yingxiang Palace

戏台正面
The stage front (photo)

当心间雀替
A *queti* (an ornate beam-supporting brace) at the central bay (photo)

当心间平身科斗栱
An intermediate *dougong* set at the central bay (photo)

当心间额枋下雕饰
A carved ornamental piece hanging down the architrave at the central bay

陕西古建筑测绘图辑

泾阳・三原

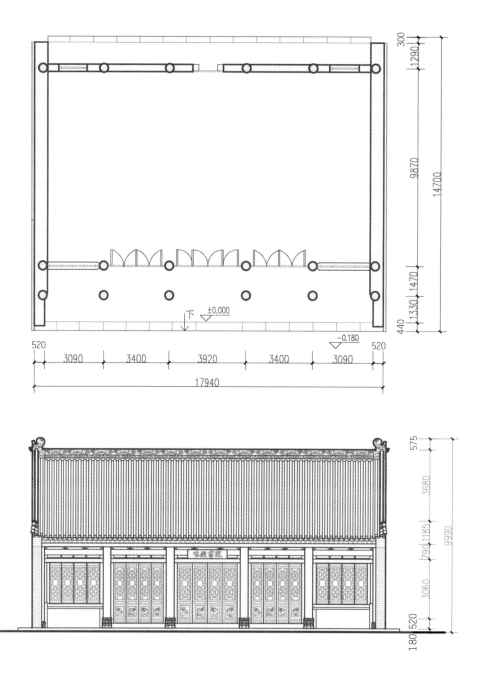

06 [清]味经书院
06 [Qing] Weijing Academy

1. 平面图
   Plan
2. 南立面图
   South Elevation
3. 现状照片
   The academy (photo)

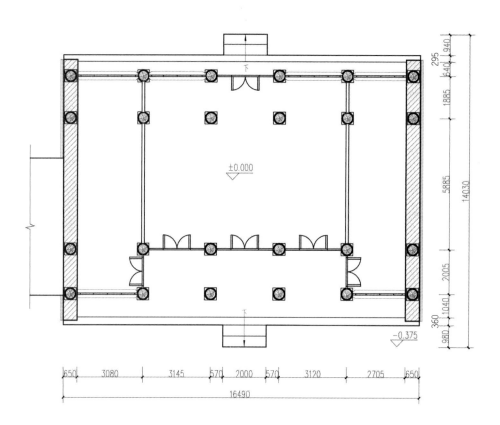

07 [清]崇实书院
07 [Qing] Chongshi Academy

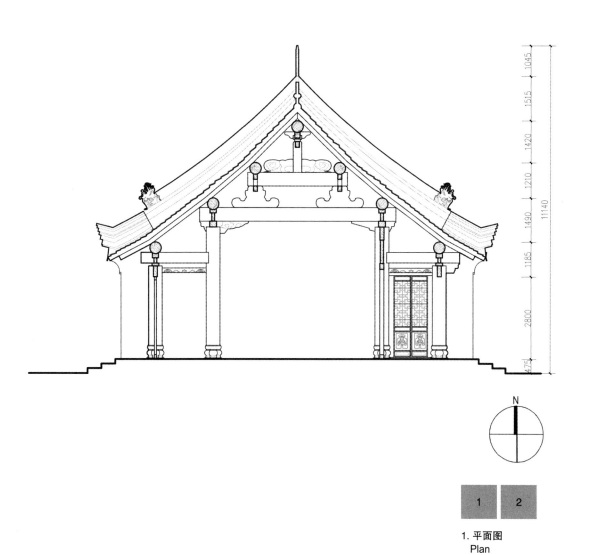

1. 平面图
   Plan
2. 横剖面图
   Transverse section

陕西古建筑测绘图辑

泾阳·三原

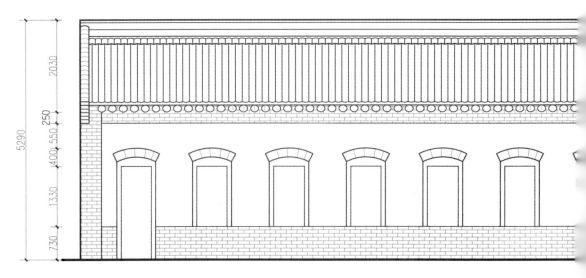

08　[中华民国] 韩家堡小学
08　[ROC] Primary School of Hanjiapu

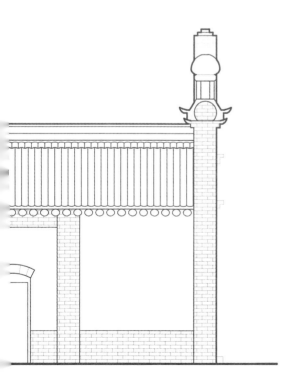

1. 东立面图
   East elevation
2. 北立面图
   North elevation
3. 东北角现状照片
   Northeast side (photo)
4. 西南角现状照片
   Southwest side (photo)

陕西古建筑测绘图辑
泾阳·三原

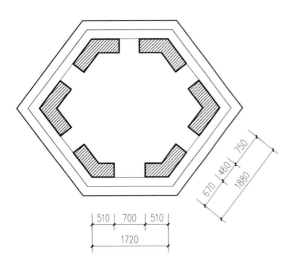

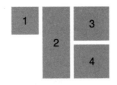

1. 平面图
   Plan
2. 南立面图
   South elevation
3-4. 二层檐部大样图
   Details, 2F eaves

09　[唐]振锡寺悟空禅师塔
09　[Tang] Zen-Master Wukong Pagoda

陕西古建筑测绘图辑
泾阳·三原

悟空禅师塔
The pagoda (photo)

09 [唐]振锡寺悟空禅师塔
09 [Tang] Zen-Master Wukong Pagoda

塔身
Upper part (photo)

塔下明代无缝塔
A tomb (photo)

砖叠涩出檐
Corbeling (photo)

檐口转角斗栱
A on-corner *dougong* set (photo)

陕西古建筑测绘图辑

泾阳·三原

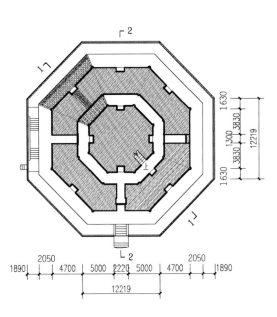

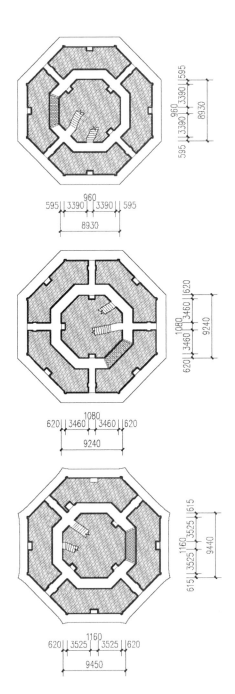

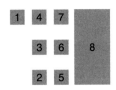

1. 一层平面图
   1F Plan
2-7. 二至七层平面图
   2F-7F Plans
8. 立面图
   Elevation

084

10 [明]崇文塔
10 [Ming] Chongwen Pagoda

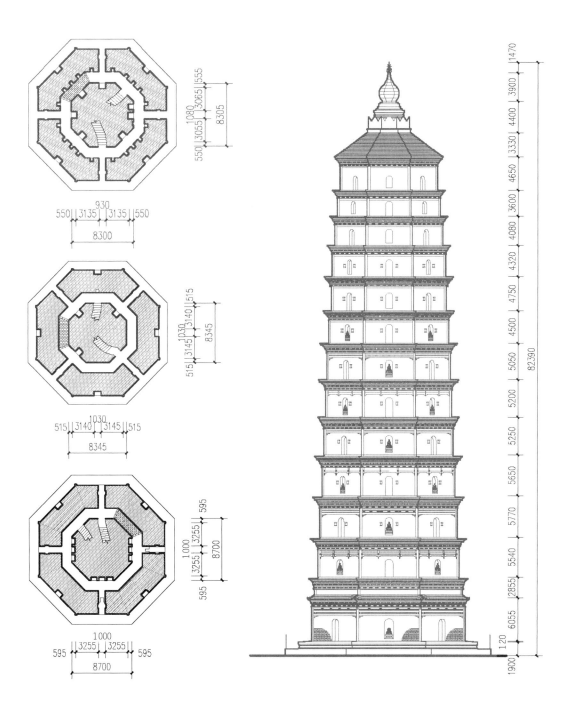

1. 大样图
   Details
2-7. 八至十三层平面图
   8F-13F Plans
8. 1-1 剖面图
   Section 1-1

10 [明]崇文塔
10 [Ming] Chongwen Pagoda

陕西古建筑测绘图辑

泾阳·三原

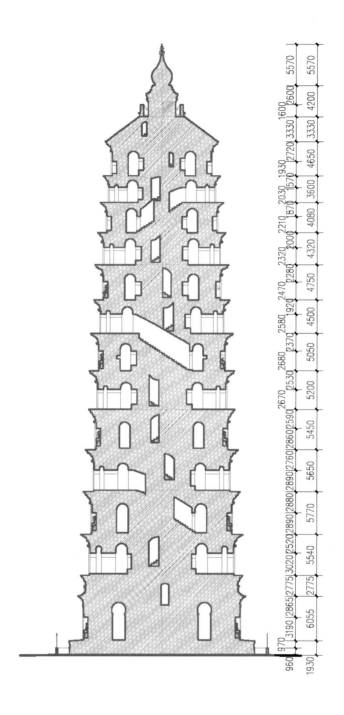

10 [明]崇文塔
10 [Ming] Chongwen Pagoda

1. 2-2 剖面图
   Section 2-2
2-5. 崇文塔现状照片
   The pagoda (photos)

陕西古建筑测绘图辑

泾阳·三原

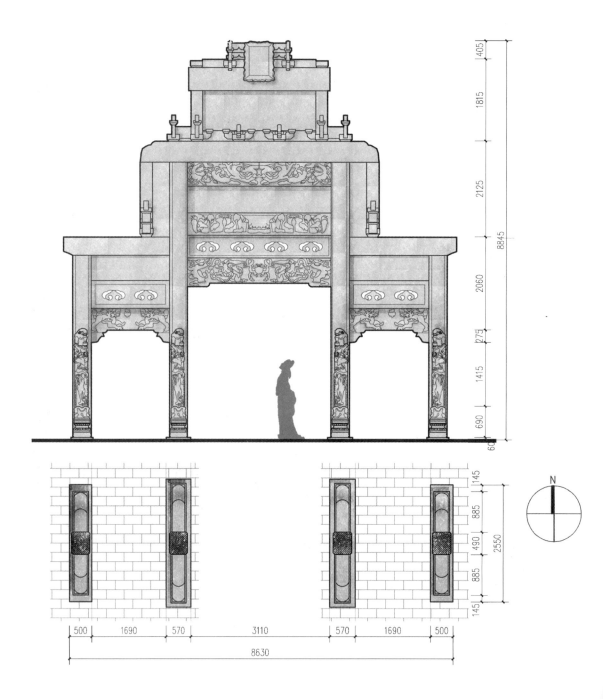

090

11 [清]张家屯村大义坊
11 [Qing] Dayi *Pailou*, Zhangjiatun Village

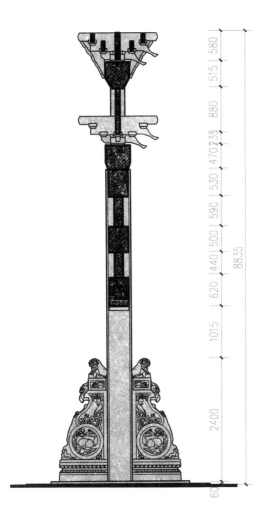
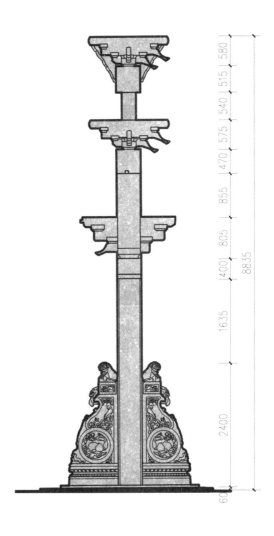

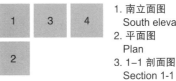

1. 南立面图
   South elevation
2. 平面图
   Plan
3. 1–1 剖面图
   Section 1-1
4. 西立面图
   West elevation

陕西古建筑测绘图辑
泾阳·三原

张家屯村大义坊
The *pailou* (photo)

顶部（北面）
Upper part, north side (photo)

顶部（南面）
Upper part, south side (photo)

11 [清] 张家屯村大义坊
11 [Qing] Dayi *Pailou*, Zhangjiatun Village

侧面
Side (photo)

横枋上浮雕
Relief on the architrave (photo)

抱鼓石上石狮
A stone lion on top of a drum-stone (photo)

抱鼓石
A drum-stone (photo)

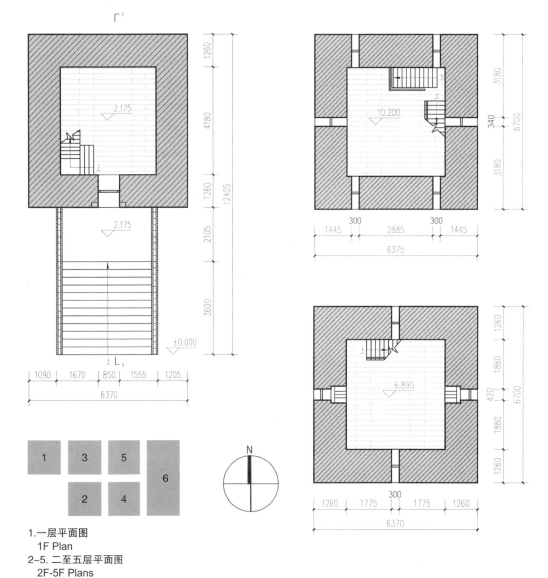

1. 一层平面图
   1F Plan
2-5. 二至五层平面图
   2F-5F Plans
6. 南立面图
   South elevation

12 [清] 李家村防卫楼

12 [Qing] Defensive Building, Lijia Village

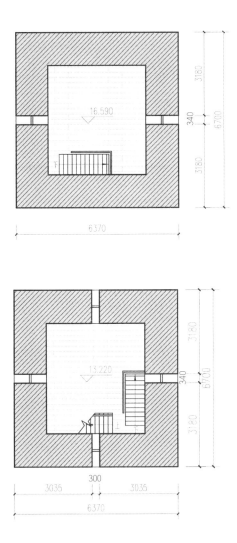
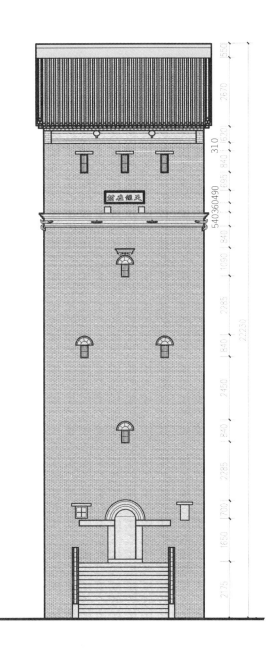

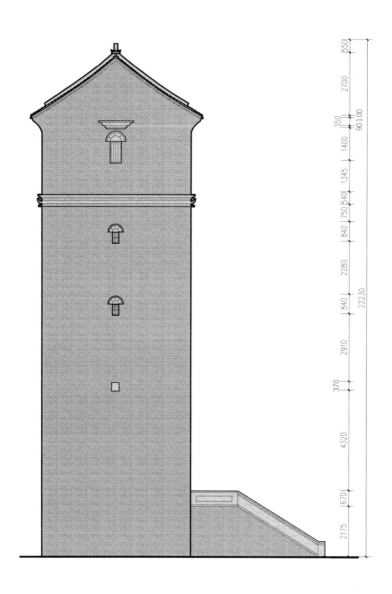

12　[清]李家村防卫楼
12　[Qing] Defensive Building, Lijia Village

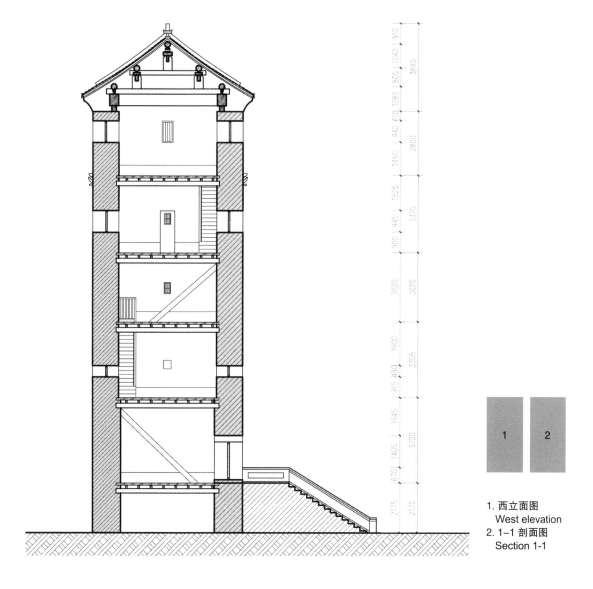

1. 西立面图
   West elevation
2. 1-1 剖面图
   Section 1-1

李家村防卫楼现状照片（西南方向）
The Defensive Building, southwest side (photo).

12 [清]李家村防卫楼
12 [Qing] Defensive Building, Lijia Village

Pillar, top of truss structure (photo)

(photo)

(photo)

李家村防卫楼南面
The Defensive Building, south side (photo)

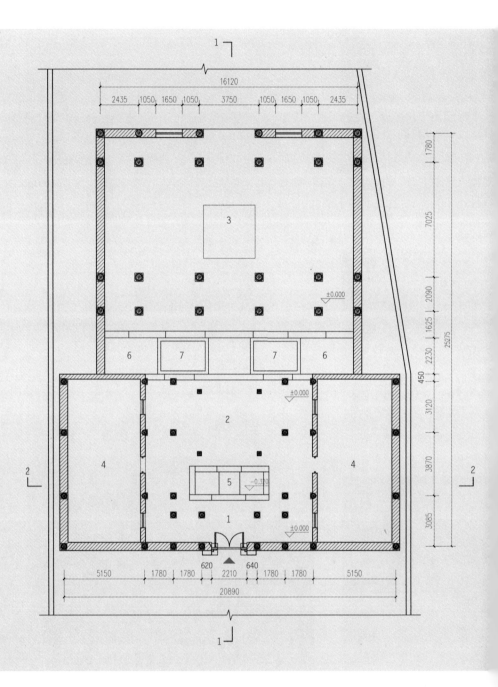

13 [清]李仪祉故居
13 [Qing] Former Residence of Li Yizhi

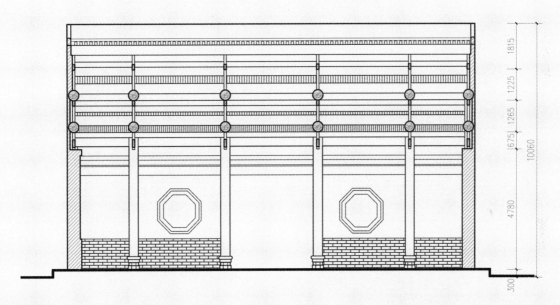

1. 门屋　　　　　5. 天井
   Gatehouse　　　Atrium
2. 戏亭　　　　　6. 内院
   Stage pavilion　Inner courtyards
3. 正厅　　　　　7. 水池
   Main hall　　　Ponds
4. 厢房
   Wings

1. 平面图
   Plan
2. 正厅纵剖面图
   Longitudinal section, the main hall

陕西古建筑测绘图辑

泾阳·三原

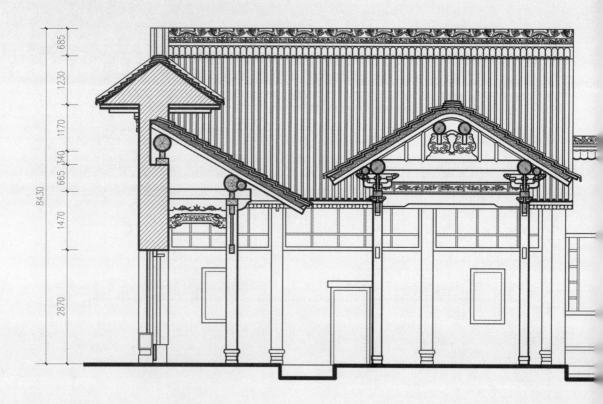

13 [清] 李仪祉故居

13 [Qing] Former Residence of Li Yizhi

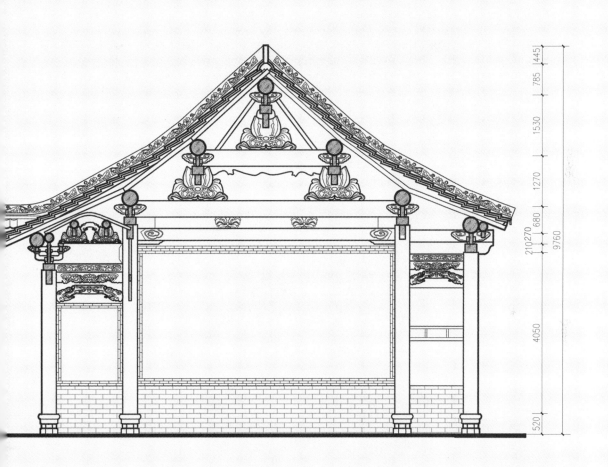

1. 1-1 剖面图
Section 1-1

陕西古建筑测绘图辑
泾阳·三原

13 [清]李仪祉故居
13 [Qing] Former Residence of Li Yizhi

1. 大样图
   Details
2. 大样照片
   Details (photos)

陕西古建筑测绘图辑
泾阳·三原

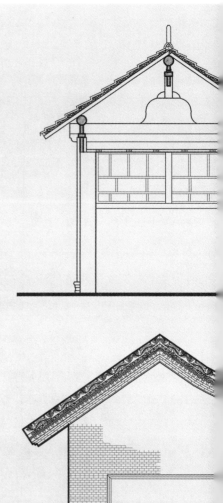

1. 2-2 剖面图
   Section 2-2
2. 入口南立面图
   South elevation, entrance

13 [清]李仪祉故居

13 [Qing] Former Residence of Li Yizhi

陕西古建筑测绘图辑

泾阳·三原

门屋
The gatehouse (photo)

院落1
East inner courtyard (photo)

院落2
East inner courtyard (photo)

13 [清]李仪祉故居
13 [Qing] Former Residence of Li Yizhi

门屋檐下斗栱
Dougong sets of the gatehouse (photo)

戏亭斗栱
A dougong set of the stage pavillion (photo)

戏亭
The stage pavilion (photo)

正厅明间梁架
The frame structure, the central bay, the main hall (photo)

内景
Interior (photo)

陕西古建筑测绘图辑
泾阳・三原

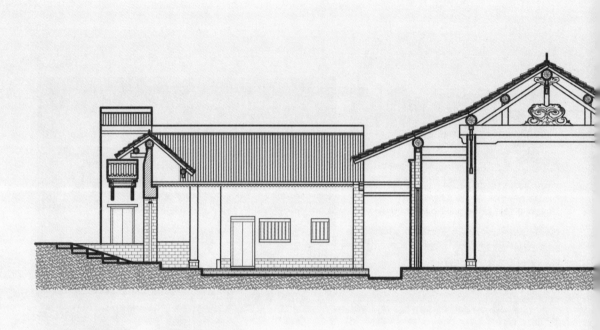

14 ［中华民国］赵春喜民居
14 [ROC] Residence of Zhao Chunxi

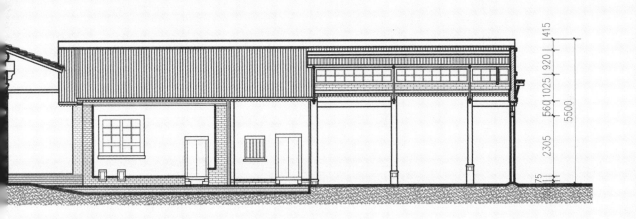

1. 院落剖面图
   Cross section, the courtyard complex
2. 现状照片
   The Residence (photo)

陕西古建筑测绘图辑

泾阳·三原

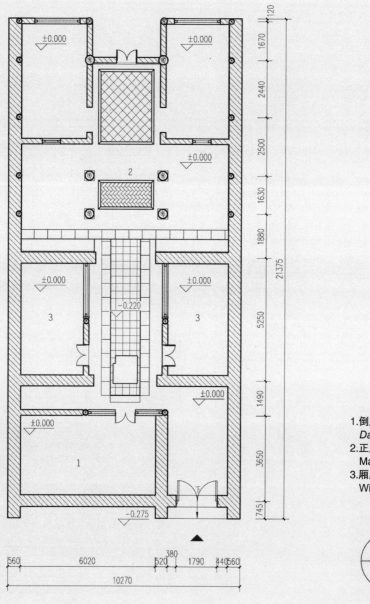

1. 倒座 *Daozuo*
2. 正厅 Main hall
3. 厢房 Wings

15 [清]蒋明杰民居
15 [Qing] Former Residence of Jiang Mingjie

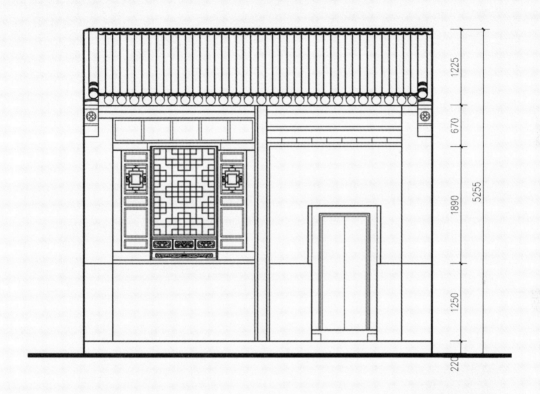

1. 总平面图
   Site plan
2. 东厢房立面图
   Elevation, the east wing

陕西古建筑测绘图辑
泾阳·三原

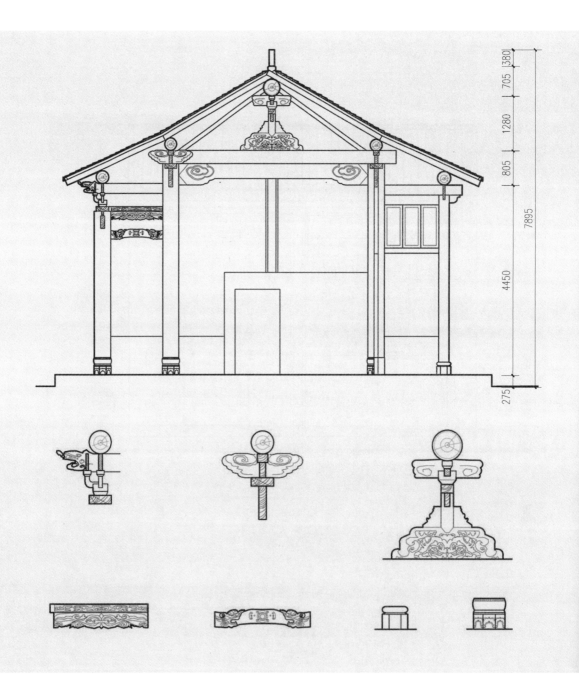

15 [清]蒋明杰民居
15 [Qing] Former Residence of Jiang Mingjie

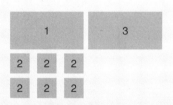

1. 正厅明间横剖面图
   Transverse section, at the central bay, the main hall
2. 大样图
   Details
3. 倒座正立面图
   Elevation, the *daozuo* (the house facing the main hall in a *siheyuan*)

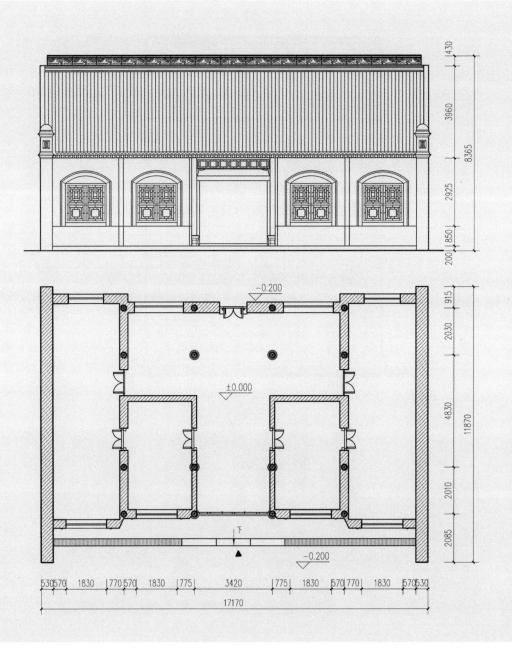

16 ［中华民国］高兰亭故居
16 [ROC] Former Residence of Gao Lanting

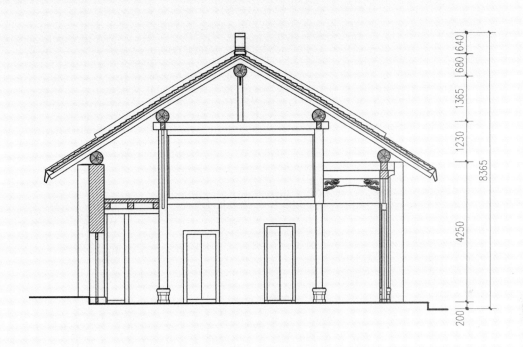

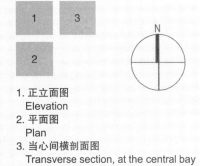

1. 正立面图
   Elevation
2. 平面图
   Plan
3. 当心间横剖面图
   Transverse section, at the central bay

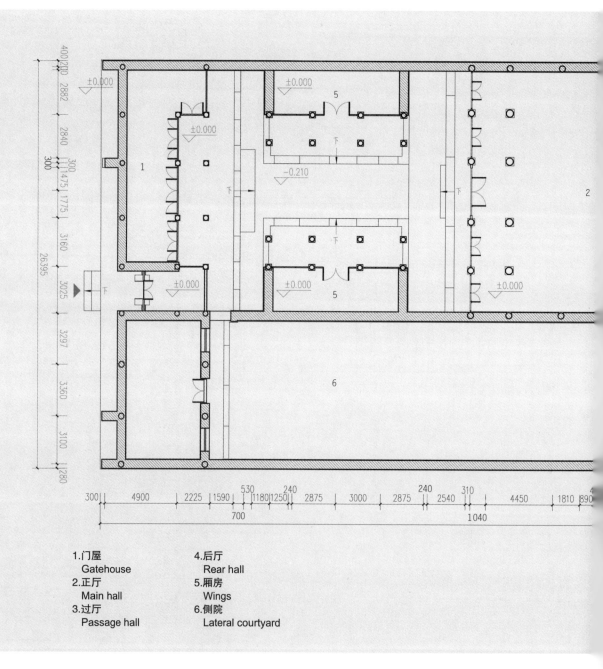

1. 门屋
   Gatehouse
2. 正厅
   Main hall
3. 过厅
   Passage hall
4. 后厅
   Rear hall
5. 厢房
   Wings
6. 侧院
   Lateral courtyard

17 ［清］吴氏庄园
17 [Qing] Wu Manor

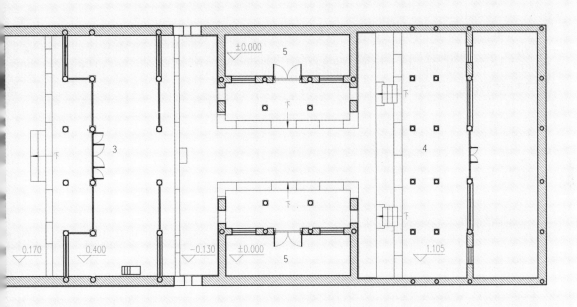

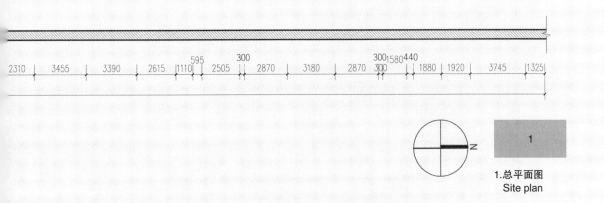

1.总平面图
Site plan

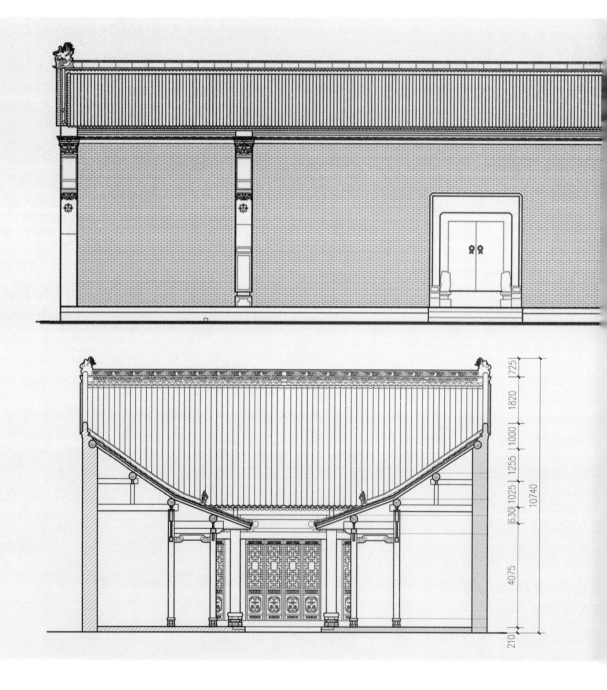

17　[清]吴氏庄园
17　[Qing] Wu Manor

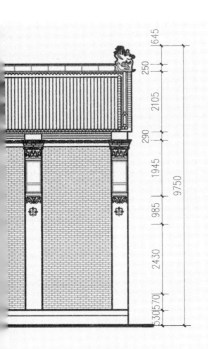

1. 门屋南立面图
   South elevation, the gatehouse
2. 正厅南立面图
   South elevation, the main hall
3. 正厅纵剖面图
   Longitudinal section, the main hall

陕西古建筑测绘图辑
泾阳·三原

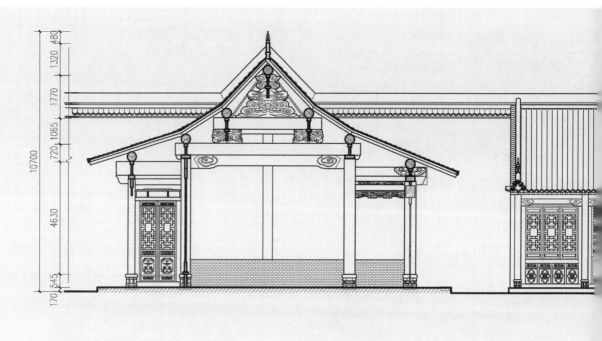

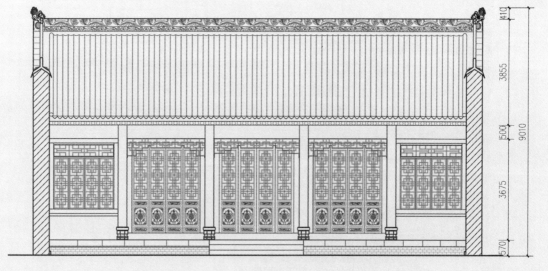

17　[清]吴氏庄园

17　[Qing] Wu Manor

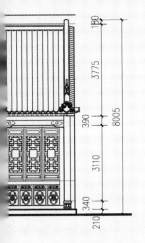

1. 正厅明间横剖面图及厢房立面图
   Transverse section at the centray bay, the main hall, plus the elevation of a wing
2. 过厅南立面图
   South elevation, the passage hall
3. 过厅纵剖面图
   Longitudinal section, the passage hall

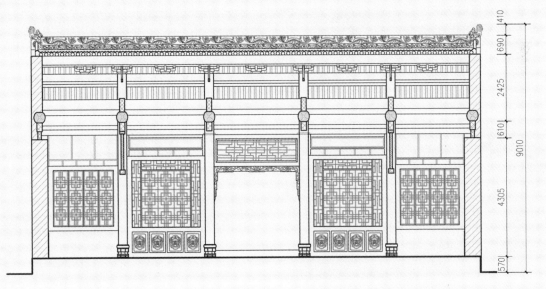

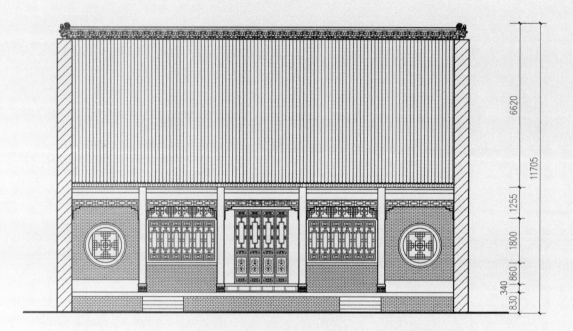

1. 后厅南立面图
   South elevation, the rear hall
2. 后厅两厢房壁门大样图
   Details, a *bimen* (an ornate gateway through the walled lateral end of the veranda)

17 [清]吴氏庄园
17 [Qing] Wu Manor

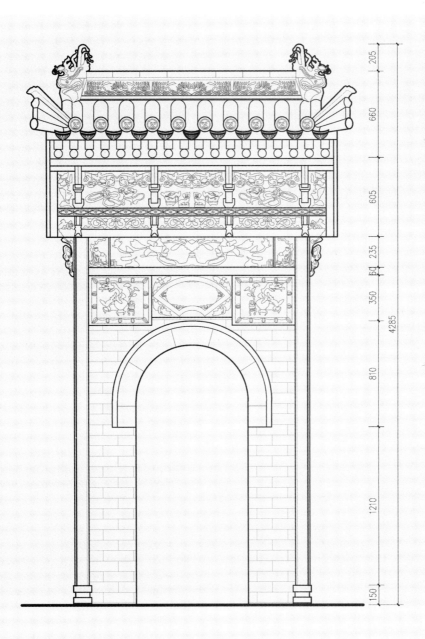

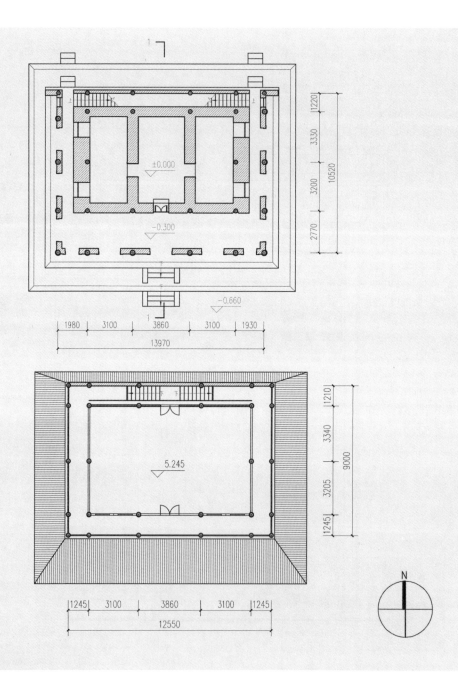

18 ［中华民国］望月楼
18 [ROC] Wangyue Building

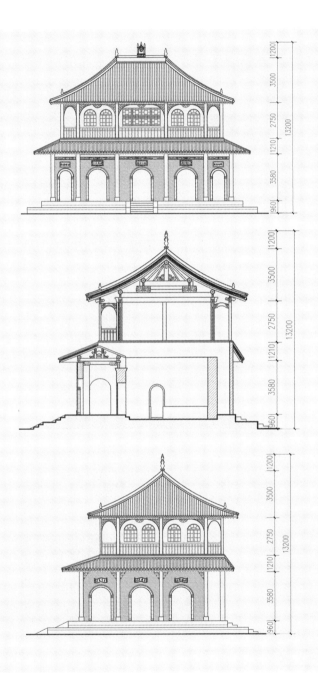

1. 一层平面图
   1F Plan
2. 二层平面图
   2F Plan
3. 南立面图
   South elevation
4. 1-1 剖面图
   Section 1-1
5. 东立面图
   East elevation

陕西古建筑测绘图辑

泾阳·三原

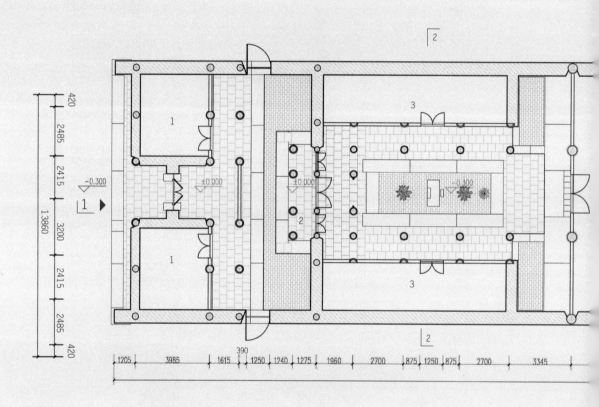

1. 门屋
   Gatehouse
2. 二门
   Secondary gateway
3. 厢房
   Wings
4. 正厅
   Main hall
5. 怀古月轩
   Huaiguyue Pavilion
6. 厢房
   Wings
7. 后厅
   Rear hall

19 [清] 孟店周宅
19 [Qing] Zhou Residence of Mengdian

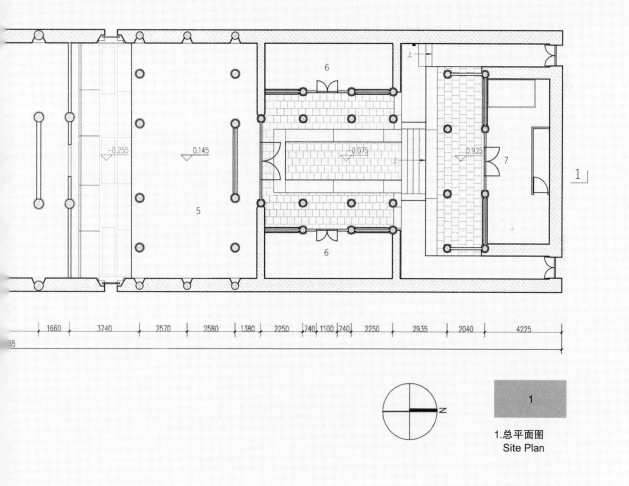

1.总平面图
Site Plan

陕西古建筑测绘图辑
泾阳·三原

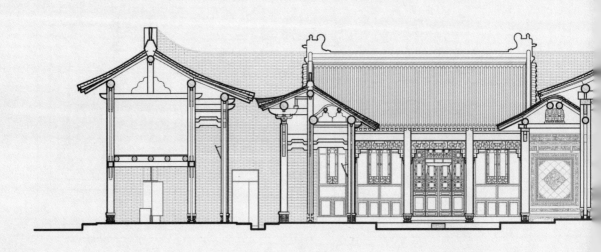

19 ［清］孟店周宅
19 [Qing] Zhou Residence of Mengdian

1. 院落1-1剖面图
Section 1-1, the courtyard complex

陕西古建筑测绘图辑

泾阳·三原

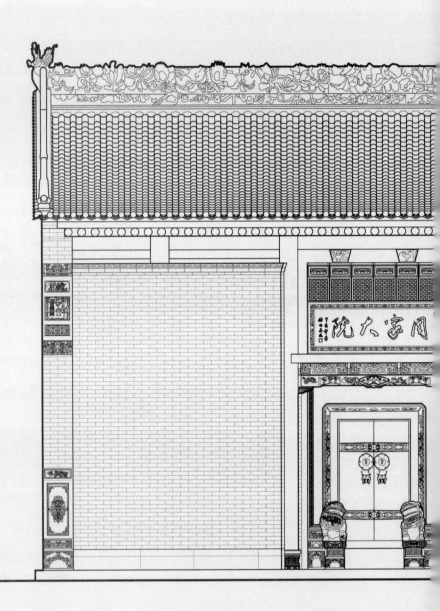

19 [清]孟店周宅
19 [Qing] Zhou Residence of Mengdian

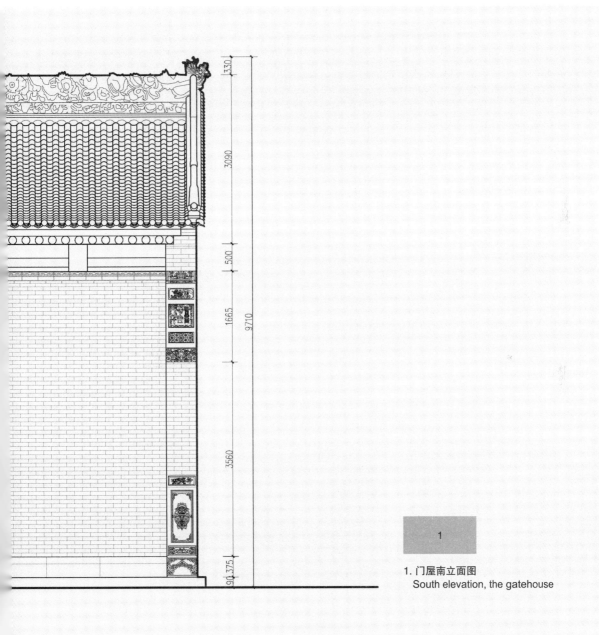

1. 门屋南立面图
South elevation, the gatehouse

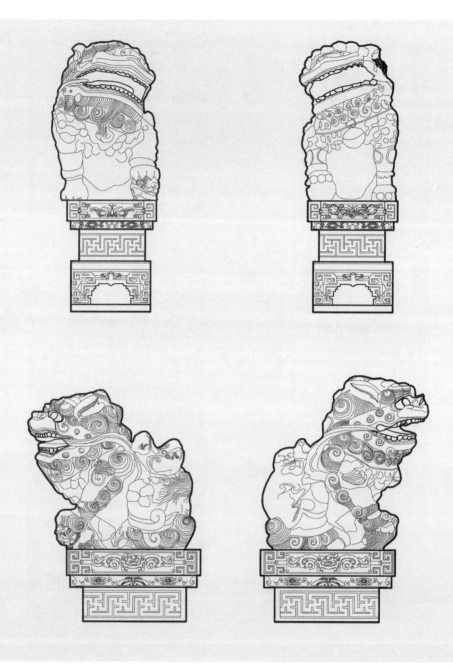

19 ［清］孟店周宅
19 [Qing] Zhou Residence of Mengdian

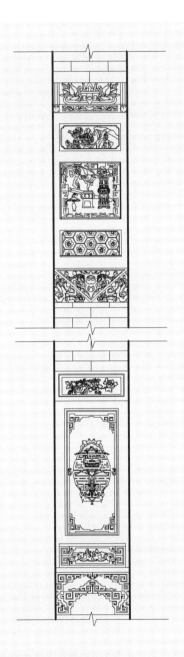

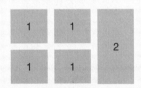

1. 门砧石狮大样图
   Details, stone lions as bearing-stones, the entrance
2. 大门砖雕大样图
   Details, brick carvings, the entrance

陕西古建筑测绘图辑

泾阳·三原

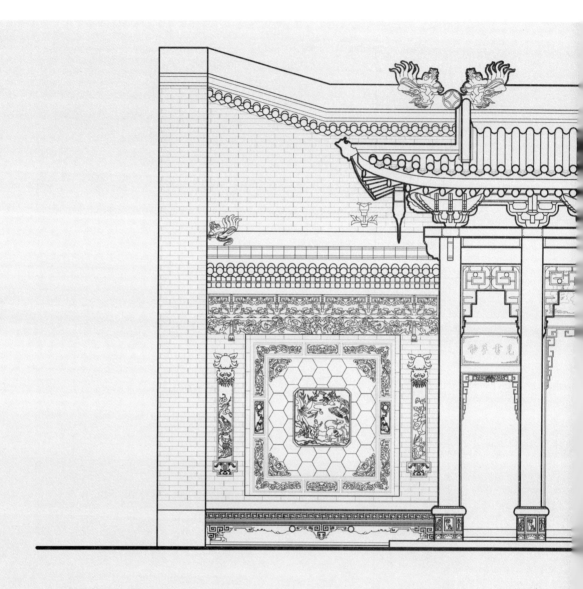

19　[清]孟店周宅
19　[Qing] Zhou Residence of Mengdian

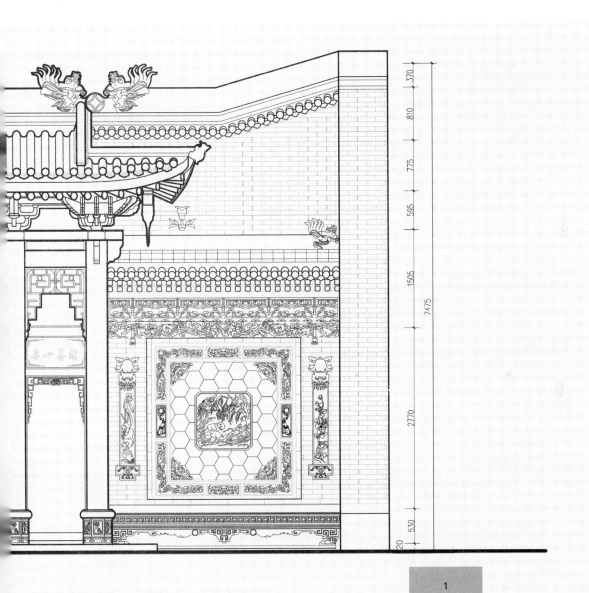

1. 二门立面图
Elevation, the secondary gateway

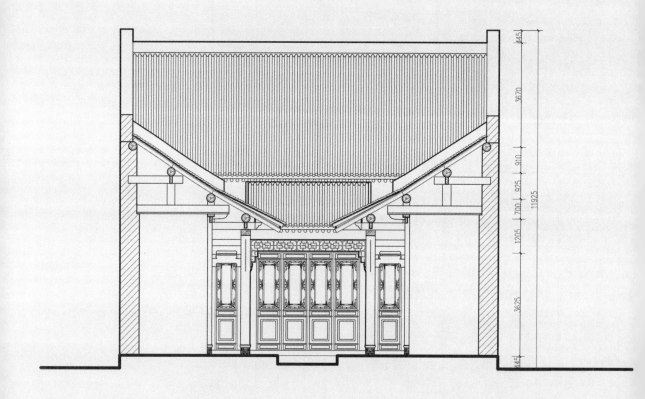

19 [清]孟店周宅
19 [Qing] Zhou Residence of Mengdian

1. 院落2-2剖面图
   Section 2-2, the courtyard complex
2. 怀古月轩南立面图
   South elevation, Huaiguyue Pavilion

陕西古建筑测绘图辑

泾阳·三原

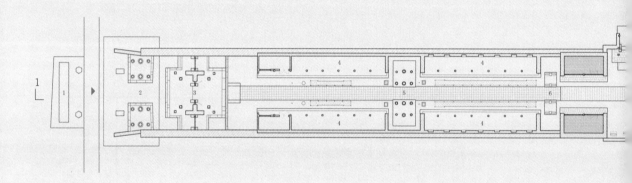

1. 照壁
   Screen wall
2. 牌坊
   Timber-structured *pailou*
3. 门楼
   Gatehouse
4. 东西碑廊
   Stele galleries
5. "陟降在兹"木牌坊
   Timber-structured *pailou* 'Zhi-Jiang-Zai-Zi'
6. "明灵保障"石牌坊
   Stone *pailou* 'Ming-Ling-Bao-Zhang'
7. 戏楼
   Stage tower
8. 东西配殿
   Wing halls
9. "明灵莫佑"木牌坊
   Timber-structured *pailou* 'Ming-Ling-Dian-You'
10. 鼓楼
    Drum-tower
11. 钟楼
    Bell-tower
12. 月台
    Platform
13. 献殿
    Offering hall
14. 正殿
    Main hall
15. 寝殿
    Resting chamber
16. 财神殿
    God-of-Wealth Hall

20 [明] 三原城隍庙
20 [Ming] Cheng-Huang Temple of Sanyuan

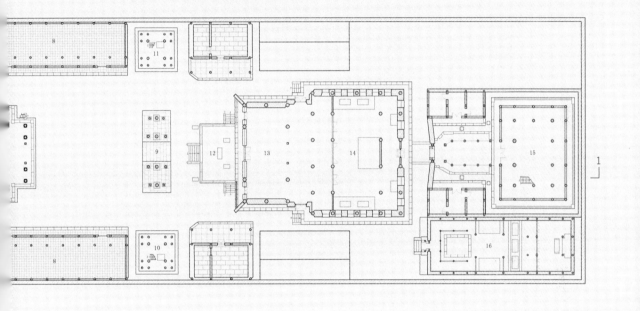

1.总平面图
Site plan

陕西古建筑测绘图辑
泾阳·三原

20 ［明］三原城隍庙
20　[Ming] Cheng-Huang Temple of Sanyuan

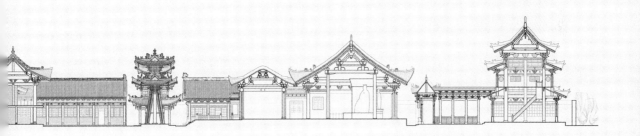

1. 院落1-1剖面图
　 Section 1-1, the courtyard complex

陕西古建筑测绘图辑

泾阳·三原

20 [明] 三原城隍庙
20 [Ming] Cheng-Huang Temple of Sanyuan

1. 照壁北立面图
North elevation, the screen wall

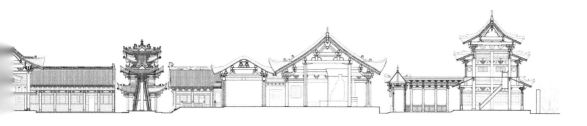

陕西古建筑测绘图辑
泾阳・三原

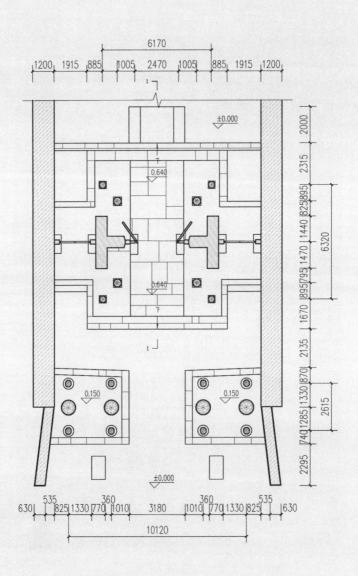

146

20 ［明］三原城隍庙
20 [Ming] Cheng-Huang Temple of Sanyuan

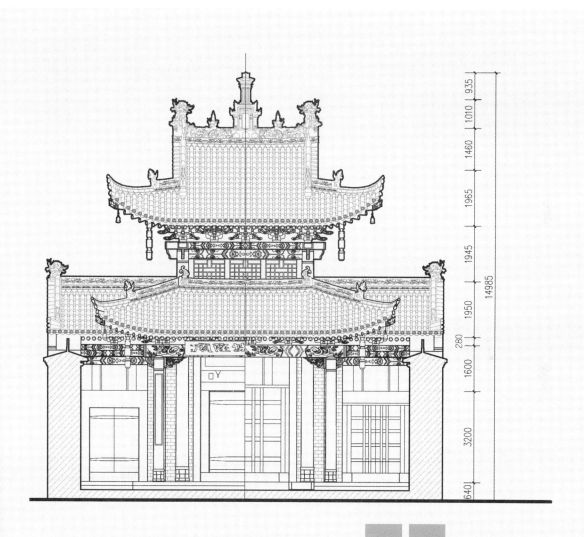

1. 门楼平面图
   Plan, the gatehouse
2. 门楼南－北立面图
   South / north elevations, the gatehouse

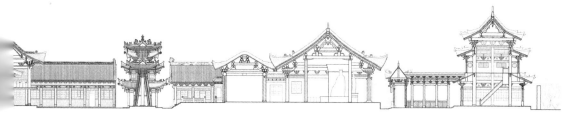

陕西古建筑测绘图辑

泾阳·三原

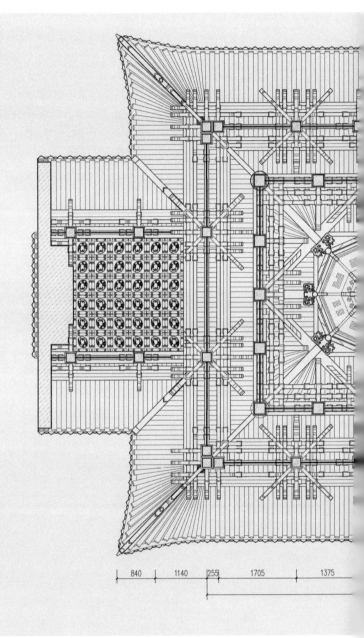

20 [明]三原城隍庙
20 [Ming] Cheng-Huang Temple of Sanyuan

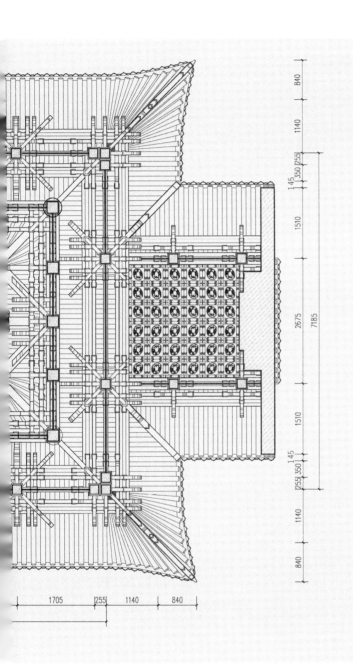

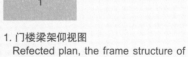

1. 门楼梁架仰视图
Refected plan, the frame structure of the gatehouse

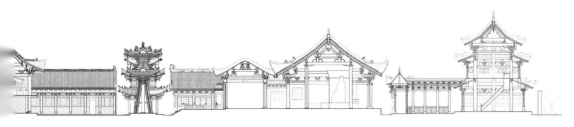

陕西古建筑测绘图辑
泾阳·三原

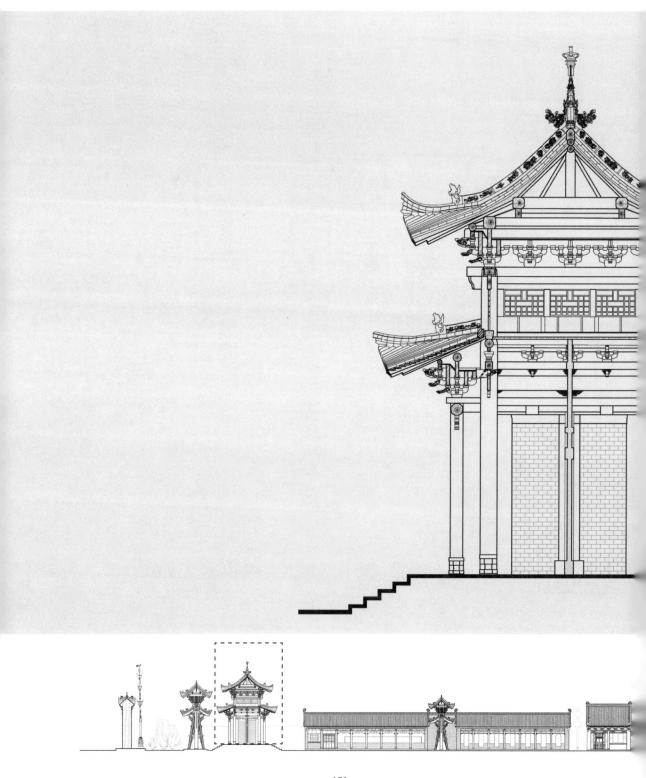

20 [明]三原城隍庙

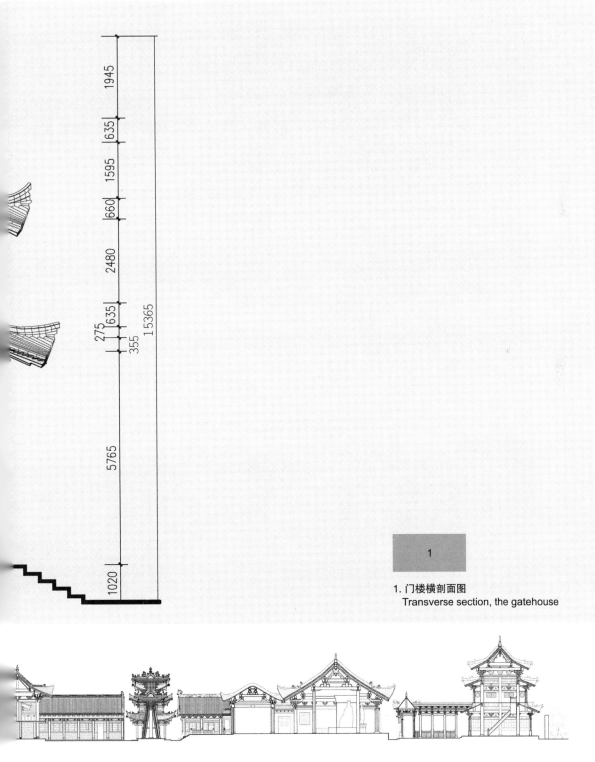

1. 门楼横剖面图
Transverse section, the gatehouse

陕西古建筑测绘图辑

泾阳·三原

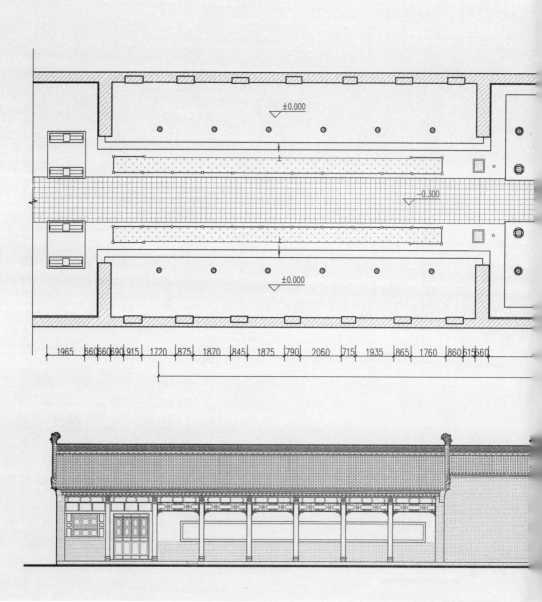

152

20 ［明］三原城隍庙
20 [Ming] Cheng-Huang Temple of Sanyuan

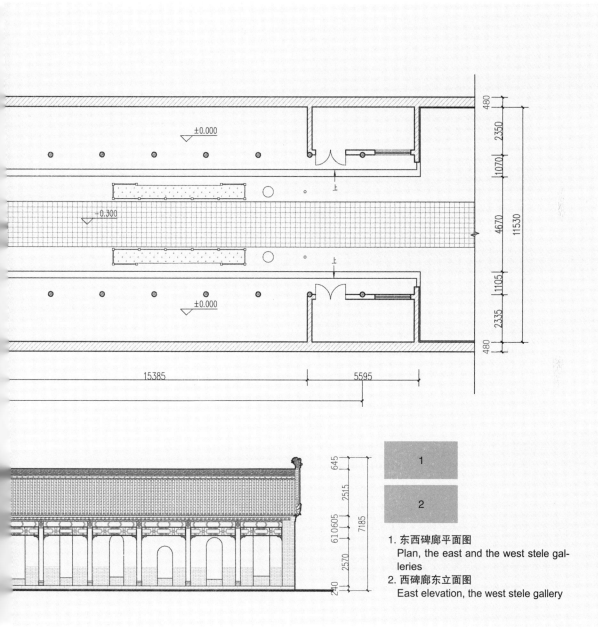

1. 东西碑廊平面图
   Plan, the east and the west stele galleries
2. 西碑廊东立面图
   East elevation, the west stele gallery

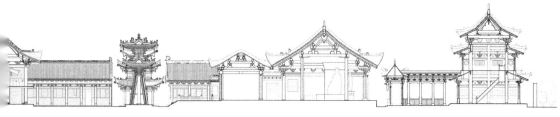

陕西古建筑测绘图辑

泾阳·三原

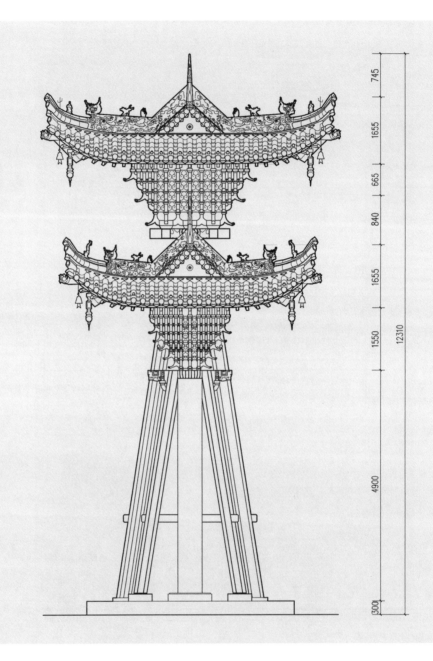

20 [明] 三原城隍庙
20 [Ming] Cheng-Huang Temple of Sanyuan

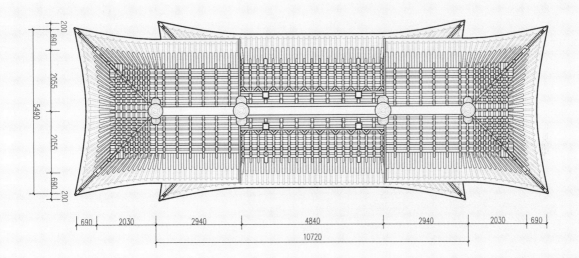

1. "明灵莫佑"木牌坊西立面图
West elevation, the 'Ming-Ling-Dian-You' (Blessed by Brilliant Spirits) timber-structured *pailou*
2. "明灵莫佑"木牌坊梁架仰视图
Reflected plan, the frame structure of the 'Ming-Ling-Dian-You' *pailou*

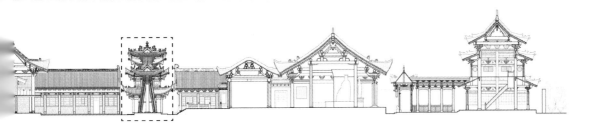

陕西古建筑测绘图辑
泾阳·三原

20 ［明］三原城隍庙
20 [Ming] Cheng–Huang Temple of Sanyuan

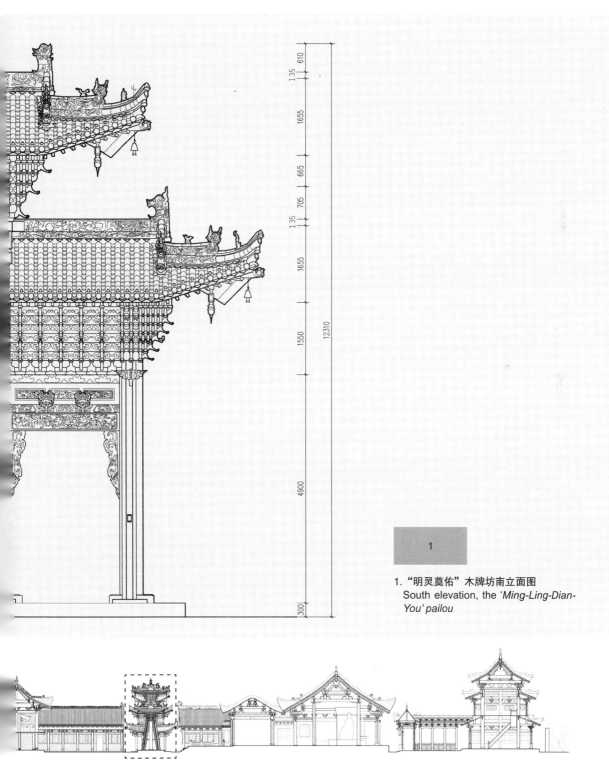

1. "明灵奠佑"木牌坊南立面图
South elevation, the 'Ming-Ling-Dian-You' pailou

陕西古建筑测绘图辑
泾阳·三原

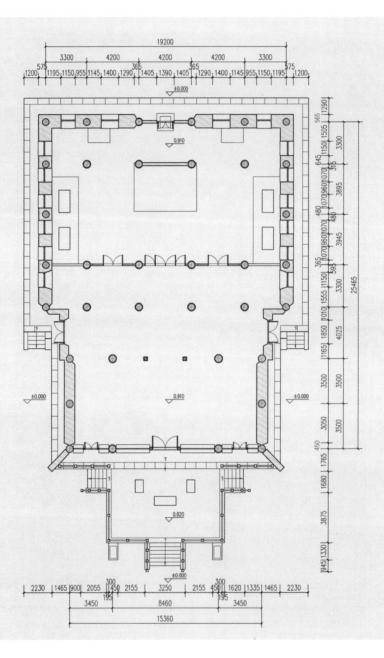
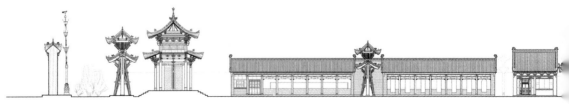

158

20 ［明］三原城隍庙
20 [Ming] Cheng-Huang Temple of Sanyuan

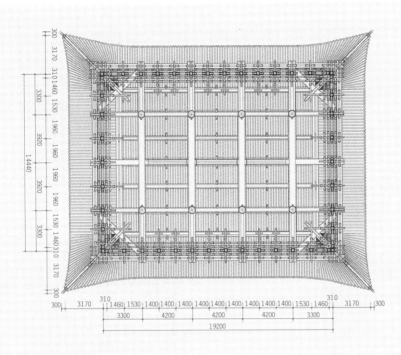

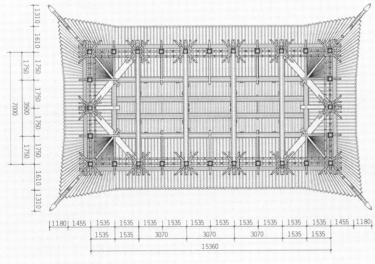

1. 大殿与献殿平面图
   Plans, the main hall and the offering hall
2. 大殿梁架仰视图
   Reflected plan, the frame structure of the main hall
3. 献殿梁架仰视图
   Reflected plan, the frame structure of the offering hall

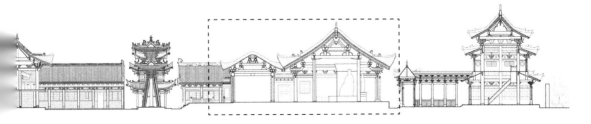

陕西古建筑测绘图辑
泾阳·三原

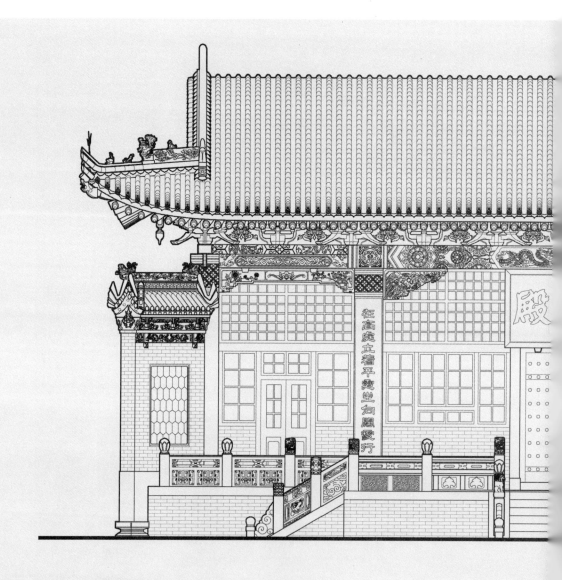

20 [明] 三原城隍庙
20 [Ming] Cheng-Huang Temple of Sanyuan

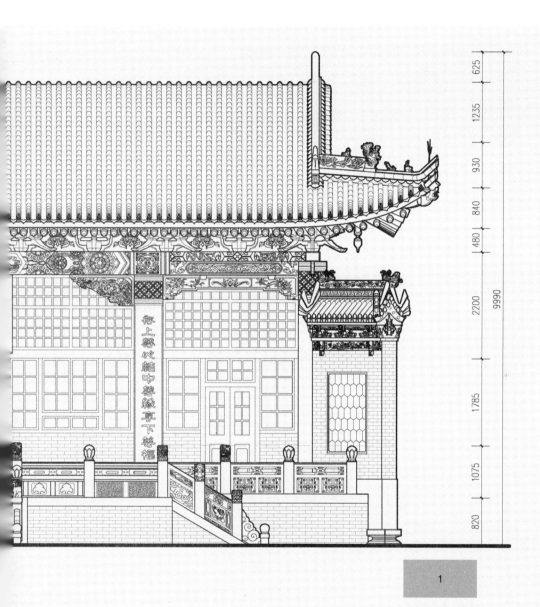

1. 献殿南立面图
South elevation, the offering hall

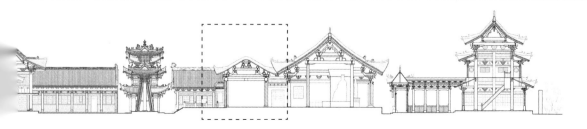

陕西古建筑测绘图辑

泾阳·三原

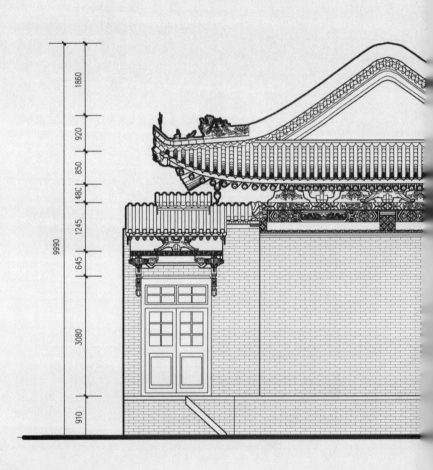

20 ［明］三原城隍庙
20 [Ming] Cheng-Huang Temple of Sanyuan

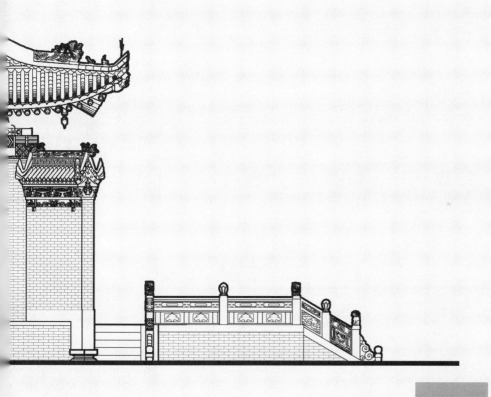

1

1. 献殿东立面图
East elevation, the offering hall

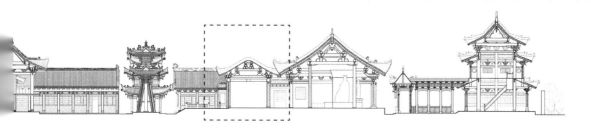

陕西古建筑测绘图辑
泾阳·三原

20　[明] 三原城隍庙
20　[Ming] Cheng-Huang Temple of Sanyuan

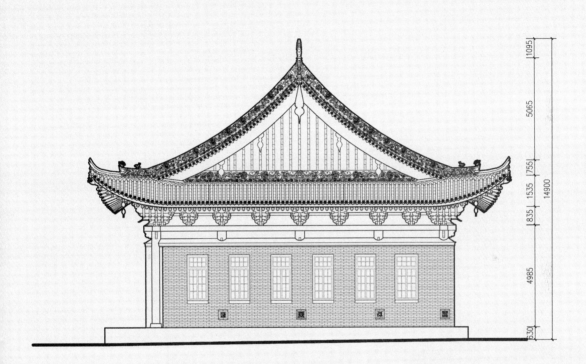

1. 大殿北立面图
   North elevation, the main hall
2. 大殿东立面图
   East elevation, the main hall

陕西古建筑测绘图辑

泾阳·三原

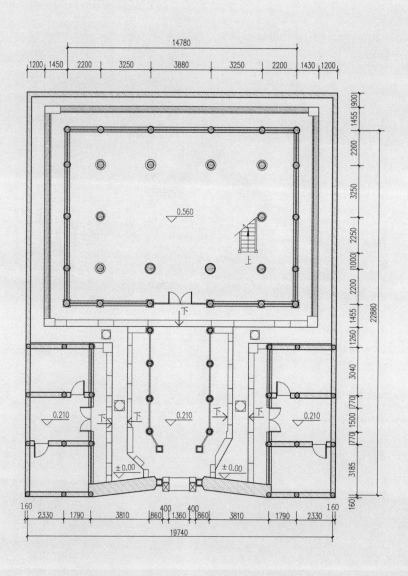

20 ［明］三原城隍庙
20　[Ming] Cheng-Huang Temple of Sanyuan

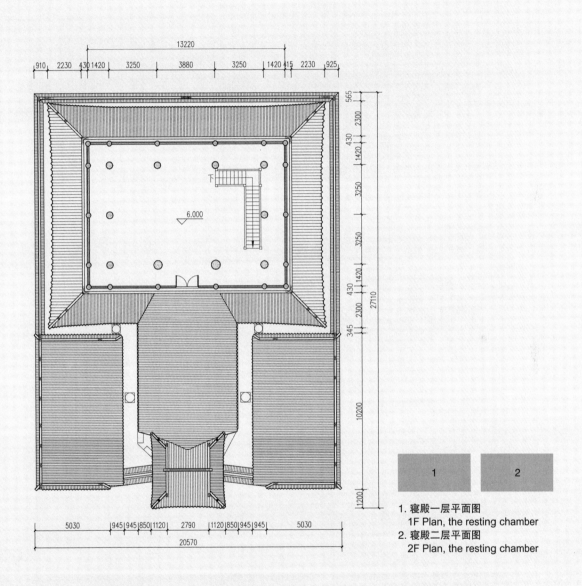

1. 寝殿一层平面图
   1F Plan, the resting chamber
2. 寝殿二层平面图
   2F Plan, the resting chamber

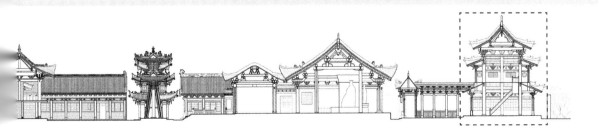

陕西古建筑测绘图辑
泾阳·三原

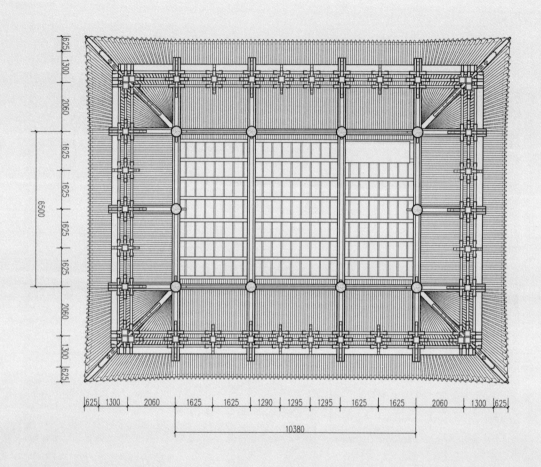

20 [明]三原城隍庙
20　[Ming] Cheng-Huang Temple of Sanyuan

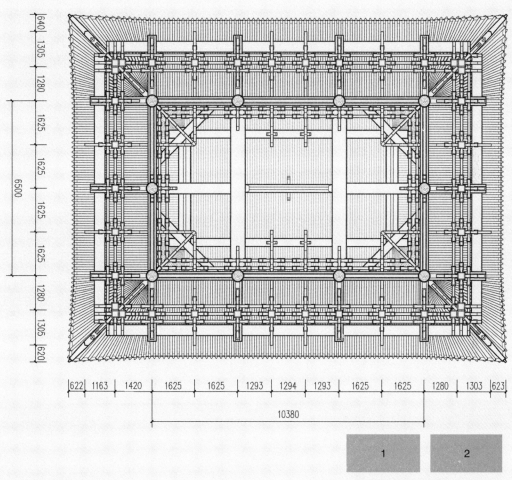

1. 寝殿一层梁架仰视图
   Reflected plan, the frame structure, 1F, the resting chamber
2. 寝殿二层梁架仰视图
   Reflected plan, the frame structure, 2F, the resting chamber

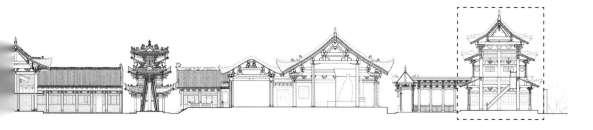

陕西古建筑测绘图辑

泾阳·三原

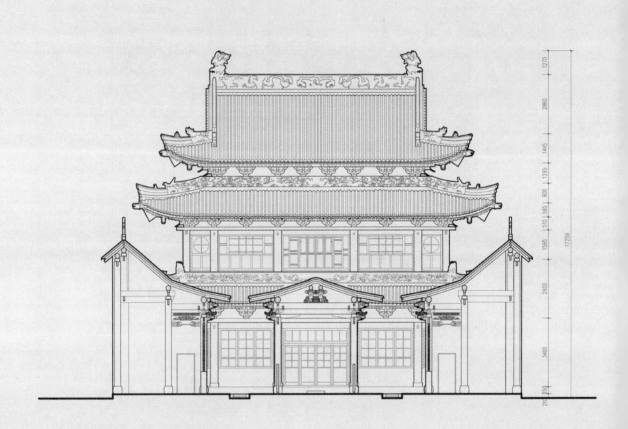

20 ［明］三原城隍庙
20 [Ming] Cheng-Huang Temple of Sanyuan

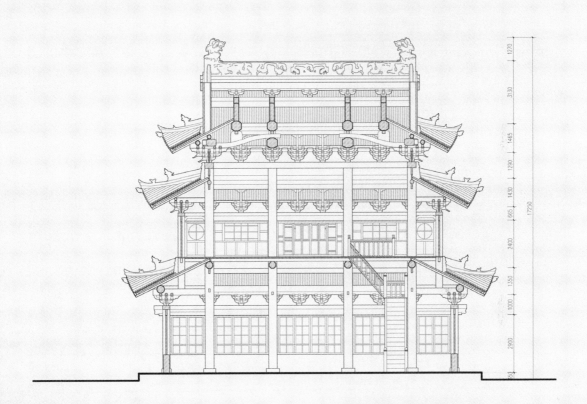

1. 寝殿院落南立面图
   South elevation, the resting chamber courtyard
2. 寝殿纵剖面图
   Longitudinal section, the resting chamber

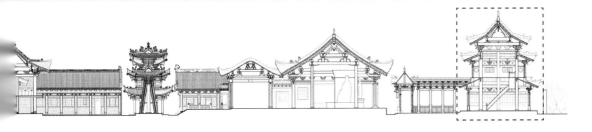

陕西古建筑测绘图辑
泾阳·三原

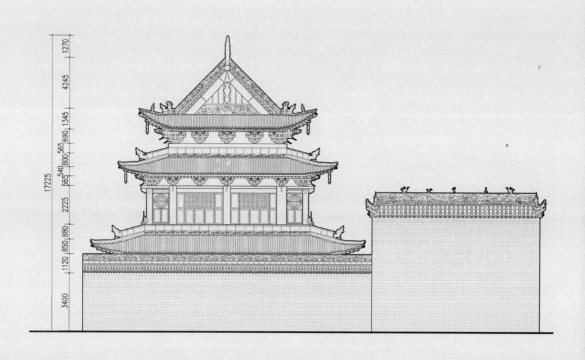

172

20 [明]三原城隍庙
20 [Ming] Cheng-Huang Temple of Sanyuan

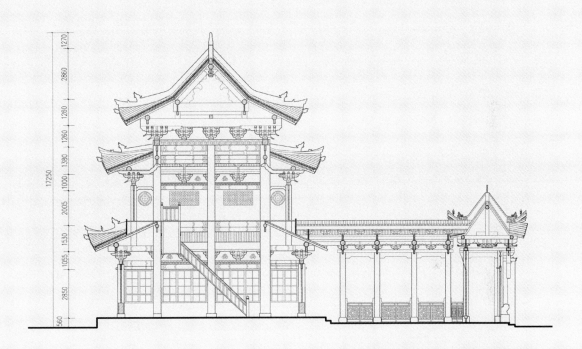

1. 寝殿东立面图
   East elevation, the resting chamber
2. 寝殿横剖面图
   Transverse section 2-2, the resting chamber

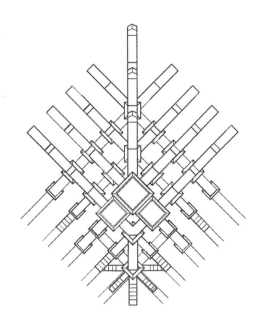

说明文字
About the Sites

01　[唐]唐德宗崇陵石刻

唐崇陵位于泾阳县安吴镇东碹村东北、嵯峨山东段南麓，唐德宗李适（742～805年，代宗李豫长子，780～805年在位）陵，陵区范围横跨泾阳、三原两县。2001年公布为全国重点文物保护单位（第五批）。

崇陵因山为陵，玄宫在嵯峨山主峰东侧，海拔955米，南向冶峪河，山环水抱。陵园占地面积约420万平方米，平面近似长方形，外围有陵墙一道（现仅剩夯土墙基），南、北陵墙走向平直，东、西陵墙则依山势修砌。陵墙四向设神门，东西二神门相距2500米，南北二神门相距1670米。神门外各立石狮一对，筑阙台两座；陵园四隅建角楼，朱雀门（南神门）外为主神道，长596米，宽71米，南端筑乳台一对。神道两侧自南而北依次排列华表、翼马、鸵鸟、仗马、石人等。

02　[唐]唐宣宗贞陵石刻

唐贞陵位于泾阳县兴隆镇崔黄村北、仲山南麓，唐宣宗李忱陵，陵区地跨今泾阳、淳化两县。2001年公布为全国重点文物保护单位（第五批）。

李忱（810～859年），宪宗第十三子，穆宗的弟弟，敬宗、文宗、武宗的叔叔，始封光王，会昌六年（846年）即皇帝位，在位十三年（847～859年）。

贞陵因仲山为陵，陵墙沿自然山势构筑，东、西二神门外两对阙址分别位于仲山东西两山的山峰之上，形势壮观。南城垣基址长1680米，北城垣4080米，东城垣2985米，西城垣4440米，内城面积约为630万平方米，几与唐太宗昭陵相当。陵墙四面辟神门，门外各置石狮一对，筑阙台一对。陵园四隅建角楼。朱雀门内有献殿遗址，门外南偏西有陵园附属建筑遗址。

朱雀门（南神门）外设主神道，长505米，宽68米。神道南端筑乳台一对，二阙基址东西间距148米。神道石刻组合与唐玄宗泰陵相同，自南而北依次为华表一对、翼马一对、鸵鸟一对、仗马三对、石人六对。

华表高约7米（石座埋入土中，未计算在内），柱身八棱，每面线刻蔓草花纹。东列华表保存较好，西列华表柱身断裂，风蚀较甚；翼马位于华表北22米处，身长2.9米，高3.3米，东列如河马状，头顶无角，尾垂，造型粗壮；西列头顶有独角，造型拙朴；鸵鸟身高1.1米，身长1.4米；仗马身高1.8米，身长2.0米，马背置鞍鞯，无马镫，披障泥，鞦秋极简，马尾下垂，脖下系铃。与其他唐陵仗马不同，贞陵仗马马背之上凿有圆窝若干；石人东列为文官，高2.5米；西列为武官，高2.9米。面容都较瘦削。

北神门外现存石马两对，东西列间距25米，形制与主神道上的石马相同。

贞陵石刻具有明显晚唐风格，雕工较为粗疏。

03 [明－清]泾阳文庙

泾阳文庙位于泾阳县城文庙街，始建年代待考。宋刻《重修文庙碑记》载有北宋哲宗元祐五年（1090年）重修的情况，可知北宋中期以前已经存在。曾毁于明嘉靖年间的关中大地震（1555~1556年），后由知县钟岱重修。明万历四年（1576年）、崇祯五年（1632年）、清乾隆二十九年（1764年）及嘉庆、道光年间均有修葺。清同治四年（1865年）知县黄傅绅修缮大成殿，光绪十一年（1885年）安吴堡吴氏庄园的周氏夫人捐银四万两再次重修。中华人民共和国成立初期泾阳文庙先后用作学校、粮站等，1980年公布为泾阳县文物保护单位，1992年公布为陕西省文物保护单位（第三批）。现用作泾阳县博物馆。

现存建筑主要有戟门、大成殿及东西庑。大成殿面阔七间，进深八架椽，单檐歇山顶；室内梁架简明、严整。梁下均用穿插枋、随梁枋垫衬；梁上用驼峰支垫檩条。外檐角科、柱头科及平身科斗栱均为五踩重昂；每间用平身科一攒，当心间平身科为三个坐斗并列，次间为两个坐斗并列，梢间只用一个，其上各出正身昂与45°斜昂，尽间平身科未用斜昂；各间平身科耍头后尾均作靴楔挑斡，挑至下金檩下；前檐柱柱础为整块青石雕成，纹饰各不相同，雕刻精美。屋面为布纹灰板瓦，黄色琉璃剪边。戟门面阔三间，进深四架椽，单檐歇山顶。每间用平身科一攒，当心间平身科为两个坐斗并列，仅当心间平身科用斜昂；屋面也为黄色琉璃剪边做法，各脊施绿色琉璃。

04 [明]太壸寺大雄殿

太壸寺大雄殿是陕西关中地区保留至今的为数不多的明代官式大木建筑，整体构架完整，保存状况良好。

太壸寺位于泾阳县城城关二条街中段，前身为十六国时期前秦苻坚（357~384年）的行宫。至隋时，文帝舍为佛寺，赐名太壸寺❶。唐玄宗天宝年间（743~756年）改为中兴寺，宋太平兴国年间（976~984年）又改为惠果寺。《陕西续通志稿》（杨虎城、邵力子，1934年编）载："惠果寺，即县内大寺，唐日本太子留学于此"。❷明景泰二年（1451年），重修大雄殿等建筑，清同治元年（1862年），寺内建筑除大雄殿外均毁于关中回乱引发的火灾。1932年大雄殿得到修缮。1957年，太壸寺大雄殿列为泾阳县文物保护单位，2003年列为陕西省文物保护单位（第四批）。

大雄殿单檐歇山顶，现状的厚砖墙和券形窗洞都是近代修缮时添建的。面阔五间，进深六架，前后乳栿用四柱，两根内柱对称布置于上金檩下，双步梁上立蜀柱承托下金檩。以前后两条大内额承托五架梁。这种利用内额减省底层大梁的做法，广泛见于元代建筑，而在明代建筑中出现是很罕有的。斗栱均为五踩重昂、扶壁重栱素方。外檐斗栱用雕刻精美的异形

---
❶ 壸，音 kǔn。《尔雅·释宫》："宫中衖谓之壸。"本意指宫中的道路，借指宫内、后宫。
❷ 此处所说日本太子或指嵯峨天皇（786-842年）。

花栱，柱头科是厢栱做成异形花栱，而平身科则是瓜栱，错落变化，甚有特色。

檐面与山面各间均用单补间，由此可知大殿的始建年代应更早，虽经明代重修仍有不少原构的早期特征保留了下来。外檐柱没有明显的侧脚与生起，立面方直、严正，呈现典型的明官式建筑风格。除早期木构特征和明官式风格之外，大雄殿的做法中同时包含的大量非官式因素体现出的则是生动的地方传统。

大雄殿内保存有北朝石立佛一尊，高度近3米，为国家一级文物。另保存有若干雕凿精美的北朝造像碑与唐《上都荐福寺临坛大戒德律师之碑》《宋游师雄题赵光辅画壁》❸碑等名碑。寺内另存有明景泰二年所立《重修泾阳惠果寺碑记》。

（关于太壶寺大雄殿的详细研究可参见：喻梦哲. 明代关中木构遗珠——泾阳太壶寺大雄殿[J]. 文物建筑，第9辑）

05 [清]迎祥宫戏台

迎祥宫戏台位于泾阳县安吴镇安吴堡村，吴氏庄园的东邻。据明嘉靖二十六年（1547年）《泾阳县志·卷二·祠庙》载，迎祥宫始建于金大定四年（1164年）。清光绪末年，慈禧避难西安时吴家的周氏夫人曾捐资十万两修缮、扩建迎祥宫。现仅存有清代戏台（兼作山门）。戏台前保存有元世祖至元十七年（1280年）所立《创建大道迎祥宫》碑一通。

戏台为2层，由戏台底层进入迎祥宫入口。戏台台口面北，南面作砖雕山门。戏台面阔三间，进深四椽；柱头科、角科、平身科均用五踩重昂斗栱，各间均用平身科一朵；当心间的平身科用三栌斗骈列，翘头和耍头均雕为动物形象。斗栱、额枋雕饰与墀头砖雕均十分精美。

06 [清]味经书院

清末陕西关中地区有著名的四大书院，即西安关中书院、三原宏道书院和泾阳的味经、崇实书院。

---

❸ 《德律师碑》现存两方，一方如《类编长安志》所记，立于西安荐福寺内；另一方则在1957年由泾川佛寺遗址迁至太壶寺保存。此碑刻于唐代宗大历六年（771年），碑身刻文16行、行36字；碑阴刻佛说弥勒成佛经，上为一佛二弟子，下左侧行书智舟法师遗记，下右侧楷书殷若波罗蜜多心经与佛说六门陀罗尼经。碑由礼部郎中韩云卿撰文、礼部尚书韩择木书丹、集贤院学士史惟则篆额、强勋刊石。
《游师雄题赵光辅画壁》碑刻于北宋哲宗绍圣元年（1094年），残存9行、行17字，原立于三原县孟店镇法相寺，1956年迁至太壶寺。碑文记载了游师雄（1038~1097年，京兆武功人，因军功累迁至直龙图阁、知秦州，喜踏古寻幽。明人赵崡称其"表章古迹，自周秦以及唐，无不有题识，至今尚存焉"）在法相寺观赏赵光辅（耀州华原人，太宗朝求学于图画院，工道释人物，兼精番马走兽，《圣朝名画评》列其番马为神品，人物为神品下。传现藏美国克里夫兰艺术博物馆的绢本《蛮王礼佛图》为其作品）壁画事，赞其"得之于心，应之于手，从容中道，左右逢源"。

味经书院在泾阳县城姚家巷中学内。2014年公布为陕西省文物保护单位（第六批）。

书院为邑绅吴建勋捐地，由陕西学政许振祎于清同治十二年（1873年）奏请修建，录取陕甘两省士子，史兆熊❹、柏景伟❺和刘光蕡❻先后主持书院。光绪九年（1883年），柏景伟主持期间订立天文、地舆、经史、掌故、理学、算学为主的教学内容，开一代新风。光绪十一年（1885年）后，刘光蕡主持，在书院内增立求友斋、通儒台（作实地测量用）、复豳馆、时务斋等，先后设立了天文、地理、算术、时政和外文等科目，使味经书院成为陕西最早讲授西学的学堂。光绪二十八年（1902年）书院停办，并入三原宏道书院。1903年改作泾阳县立小学堂，1923年更名泾阳县高等小学，今为姚家巷中学。

书院原有大门、照壁、牌坊两座（东书"英俊之域"，西书"礼乐所流"）、舍房、斋房、讲堂、东西庑、藏书楼、通儒台及清白池等，中心建筑为讲堂。院落总长度约170米，宽度近40米。现仅存讲堂，面阔五间，进深六架椽，带前廊，硬山顶。

07 [清]崇实书院

崇实书院是清末陕西关中地区著名的四大书院之一。2014年公布为陕西省文物保护单位（第六批）。位于县城姚家巷县委大院内。

崇实书院是以味经书院的时务斋为基础，于清光绪二十三年（1897年）建立的，设"政事"、"工艺"二斋，"政事"主要是了解各国现状，探求救亡之法；"工艺"则是仿制器具，推广实业。由此将书院命名为"崇实"。清光绪二十八年（1902年）书院停办，拆除部分房屋。宣统二年（1910年）改为泾阳县工业学堂。现仅存清代讲堂一座。坐北朝南，面阔五间，进深六架椽，硬山顶。

08 [中华民国]韩家堡小学

韩家堡小学位于安吴镇韩家堡村，是陕西地区现存为数不多的民国时期的教育建筑，现仅存校舍一座，长方形平面，正门开设在东山墙面上。已荒废不用。

---

❹ 史兆熊（？~约1883年），字梦轩，陕西城固县人。泾阳味经书院首任院长。

❺ 柏景伟（1830~1889年），字子俊，号沣西老农，陕西长安人。教育家。先后在味经书院、泾干书院、关中书院讲学。主持味经书院期间创立了书院的刊书处，刊刻了许多新书及科技著作。经他倡导，在陕西设立了官办书局。还曾创立长安牛痘局，为当地儿童接种牛痘。光绪十一年（1885年），受陕西学使之约移讲关中书院并主持书院。著有《沣西草堂集》、《柏沣西先生遗集》等。

❻ 刘光蕡（1843~1903年），字焕唐，号古愚，陕西咸阳人。思想家、教育家，清末维新运动的西北地区领袖，与康有为并称"南康北刘"（康有为称他是"海内耆儒"，梁启超称他为"关学后镇"）。曾任泾阳味经书院院长，创办泾阳崇实书院，在咸阳、扶风、礼泉等地创办义学多处，后应陕甘总督之邀赴甘肃任甘肃大讲堂总教习。培养了众多具有近代科学知识的人才，著名的"关中三杰"——于右任、李仪祉、张季鸾皆出其下。晚年居礼泉九嵕山山下烟霞草堂讲学。著有《烟霞草堂文集》等。

09 [唐]振锡寺悟空禅师塔

振锡寺悟空禅师塔位于泾阳县嵯峨山第二峰峰顶。2008 年公布为陕西省文物保护单位（第五批）。

悟空禅师，京兆云阳（今属泾阳县）人，俗姓车，名奉朝，生于唐玄宗开元二十年（735 年），圆寂于唐宪宗元和七十年（812 年），本为鲜卑拓跋氏后裔。唐玄宗天宝十年（751 年），罽宾国酋领摇婆达擀和三藏舍利越摩来朝，唐派遣使臣回访，车奉朝随行出使，期间因病居于犍陀罗国，拜舍利越摩为师，法号达摩驮都，参与翻译了多部佛经，于德宗贞元五年（789 年）返回长安，改法号为悟空，住长安章敬寺。圆寂后归葬故乡嵯峨山。宪宗敕建振锡寺以资纪念❼，后于唐宣宗大中十四年（860 年）建塔（据《唐悟空禅师塔》塔铭载）。塔下散置有明舍利塔（无缝塔）一座，塔身刻有《大明新建嵯峨山中五台振锡寺》塔铭，款为"嘉靖二十年三月望日僧人如通立"，由此可知振锡寺曾于明嘉靖二十年（1541 年）重修，但不知是否同时也对塔进行了重修。

悟空禅师塔为楼阁式实心砖石塔，平面为正六边形，底层边长 1.8 米；5 层，存高 13.3 米。塔基、底层和第二层塔身的下半部为块石砌筑，第二层上半部以上为青砖砌筑；底层各面均做券龛。第二、三层南北面对开券门，其余各面做券龛。龛内置佛像，现均已失。第三层以上塔身高度骤减，各面均为素面，无券门或券龛；塔的底层雕出阑额与斗栱及圆形倚柱（仿木构砖壁包柱做法）。柱头上为单杪四铺作斗栱；第二、三各层塔身均用平砖隐砌出倚柱、阑额和斗栱。阑额上承单杪四铺作柱头斗栱一朵与转角斗栱各一朵；塔的底层用块石雕出檐部，第二层以上用平砖叠涩出檐。塔的保存状况不佳，塔身开裂且倾斜严重，顶部残毁，植物生长茂盛，塔刹已不存。

10 [明]崇文塔

崇文塔位于泾阳县崇文镇东太平村、泾水北岸。是我国现存第二高的砖塔。塔建于明万历年间，为倡导泾阳、三原和高陵三地之文风而建，故名"崇文"。崇文塔下原本有铁佛寺，清同治元年（1862 年）毁于关中回乱，仅塔保存下来。1956 年公布为陕西省文物保护单位（第一批），2001 年公布为全国重点文物保护单位（第五批）。

崇文塔为楼阁式空心砖塔，平面为正八边形，底层边长 9.3 米，13 层，通高 87.2 米。塔底层南面辟券门，门额题"崇文宝塔"四字。其余各面均辟一券龛，内置佛像一尊；第二层以上每层各面均相间辟券门和券龛，上下层之间券门和券龛位置也是交错设置，龛内均有佛像 1 尊（个别佛像已遗失）。塔壁上部用平砖隐砌倚柱、额枋和斗栱；塔壁角部均隐砌圆柱，

---

❼ [宋] 宋敏求，《长安志·卷二十》："振锡寺在县北一十五里嵯峨山上，唐元和七年置。"

第二、四、六层塔壁增设垂莲柱，将壁面分为三间；额枋上雕荷花、瑞兽、飞天、寿星等图案；斗栱为坐斗加麻叶出卷云头式，补间斗栱数量自下而上逐层减少，底层九朵、第二~七层八朵、第八~九层七朵、第十~十三层六朵，柱头上转角斗栱各一朵。各层间叠涩出檐，檐口两排仿木方形椽头，塔檐上面形成宽约1米的腰檐平台，可从券门出至此平台。

塔内有塔心柱，塔壁与塔心柱之间为券顶式环廊，楼梯沿塔内壁和塔心柱盘旋而上，可至塔顶。塔顶为平砖叠涩，上置4.2米高铜质宝瓶式塔刹，周围有护墙，可登顶远眺。

[附：明万历十九年（1591年）重修铁佛寺时邑人李念慈所撰《铁佛崇文塔寺常住田供衣记》碑碑文：

"县东旧有铁佛寺，在今寺北，年久倾坏，其移而南，建有塔也。则自我五世祖太保谥敏肃公、□□大王父民部公、泊王父中翰公，踵修塔十有三级。太保公及世成七级，民部公九级，越中翰公竣。初，了凡袁先生游秦，太保公延授民部公。尚书先生故善相地，因相塔处曰：是泾、原、高陵三邑文章节义之风气实赖焉。乃命今名。经营于万历十有九年，至三十有六年，凡十八年而毕。寺基广七十二亩，塔基广亩有七分，高三十六丈，中空。自颠垂□及泉，上下安舍利七颗……"]

11 [清]张家屯村大义坊

张家屯村大义坊位于泾阳县云阳镇张家屯村村口大道正中。

大义坊是一座四柱三间五楼重檐石牌坊。底层东西宽2.74米，南北长8.34米，通高8.78米。檐下用单昂斗栱，顶部严重倾颓，部分构件已斑驳脱落。石牌坊上雕刻有龙、鹤、虎、狮等形象。背面立柱上刻有"敕赠儒林郎南京寺珍馐署署正张巍"字样，推测为明代皇室赐予执掌御膳的官员张巍的。

12 [清]李家村防卫楼

李家村防卫楼位于泾阳县安吴镇西李家村十字大道南，是一座建于高大台基上的砖楼，5层，通高22米。坐南面北，底层东西8.75米，南北6.7米。内设木梯，各层均设有券窗或瞭望孔。墙体厚度逐层递减，四、五层之间有出檐。五层雕有"天鉴在兹"四字匾额。

李家村防卫楼是为抵御土匪而造，据传由唐卫国公李靖的后代创建，清末重修，是陕西地区现存数量非常有限的建筑类型，2008年公布为陕西省文物保护单位（第五批）。

13 ［清］李仪祉故居

李仪祉故居（中华水利会馆）位于泾阳县王桥镇社树村。现存一进院落，坐南朝北，有门楼、东西厢房、戏亭和正厅。整个建筑轩敞，雕梁画栋。砖雕门楼檐下砖雕精美，浮雕45°斜出栱，其上是随栱斜出的耍头。院落中有戏亭一座，为两椽卷棚，雕刻甚为华美。正厅面阔五间，进深六架椽，前廊步下用茶壶轩，颇有南方民居建筑的意趣。每根檩条下均以翼栱穿插垫板，增进装饰的同时也加强了关键节点的稳定性；脊檩与上金檩下均用高驼峰，雕饰精致细腻。

李仪祉❽先生任陕西省水利局局长期间，在此设立泾惠渠工程指挥部并工作多年，主持修建了泾惠渠这一惠及关中百姓的重大工程。李仪祉故居于2014年列为陕西省文物保护单位（第六批）。现已荒芜废弃。

14 ［中华民国］赵春喜民居

赵春喜民居位于泾阳县口镇。口镇曾是通往陕甘宁边区的交通要津。院落保存完整，一进，由门屋、东西厢房和正厅组成。门屋前檐短后檐长，这种长短坡屋顶是此地区民居建筑的一个普遍特点；正厅面阔三间，进深六架椽，雕饰比较精美。其院落布局与构造做法较为典型地体现了本地区民国时期普通民居建筑的特点。

15 ［清］蒋明杰民居

蒋明杰民居位于泾阳县安吴镇蒋路村，始建于清中期。坐南朝北，现仅存第一进院落，有门屋、东西厢房和正厅。东、西厢房为硬山顶，两开间；东厢房的南山墙作影壁，雕有几何花卉纹饰；正厅面阔三间，进深四架椽，前后内柱柱头用翼形栱绞金垫板，上承金檩；驼峰和穿插枋、随梁枋的端头均有精美雕饰。正厅的做法前繁后简，同类构件前檐的做法明显比后檐细致、考究，例如前檐用斗栱，后檐则直接以柱子承托垫板；前檐单步梁整体做雕饰，出头部分雕作龙头，而后檐的单步梁则无雕饰。蒋明杰民居虽然院落保存已不完整，但是现存的几座单体建筑都基本完好，是关中地区民居建筑的典型代表。

---

❽ 李仪祉（1882～1938年），陕西蒲城人，著名水利学家和教育家，我国现代水利建设的先驱。考取秀才后入崇实书院学习西学，后入京师大学堂学习并留学德国。亲自主持建设陕西泾、渭、洛、梅四大惠渠，是我国现代水利工程的典范，对我国水利事业做出重大贡献。还创办了我国第一所水利工程高等学府——南京河海工程专门学校。曾任国立西北大学校长，并在北京大学、清华大学、同济大学等多所大学执教。

16 ［中华民国］高兰亭故居

高兰亭❾故居位于泾阳县城二条街东段，属于清末民初泾阳县城的城市住宅。现存门屋、厢房和正厅，其中正厅经过改造，已非原貌。

17 ［清］吴氏庄园

吴氏庄园位于泾阳县安吴镇安吴堡村。1991年公布为陕西省文物保护单位（第三批），2013年公布为全国重点文物保护单位（第七批）。

吴氏庄园由盐商吴尉文在清末始建。其子娶三原县鲁桥镇孟店村周氏（1868～1910年）为妻，病故后因无子嗣遂由周氏管理家族生意。清光绪二十六年（1900年），八国联军攻占北京，慈禧太后逃至西安避难，周氏贡银十万两，诰封为二品夫人，慈禧还赐亲笔书"护国夫人"金字牌匾。周氏日常亦为赈济灾民、修建文庙、书局印书等公益事出资捐助，使吴家成为关中著名的大家望族。

吴氏庄园坐北面南，南北总长约70米，东西近30米，现存三进院落，中轴线上依次为门厅、过厅、正厅和后厅。其中门厅为东、西两座倒座拼成，通面阔八间，进深三架椽（前檐长后檐短）；过厅面阔五间，进深三架椽（前檐短后檐长）；正厅面阔五间，进深六架椽；后厅面阔五间，进深四架椽（前檐长后檐短）。长短坡屋面的做法，是陕西地区民居建筑的常见做法，可在经济合理的使用面积内，尽可能多地获取充分的光照。柱础、墙面等处的石雕、砖雕甚是精致考究。最后一进院落西厢房南山墙上的砖雕尤为精美，是关中地区民居砖雕中的上乘之作。

抗日战争期间，中共中央青年工作委员会领导西北青年救国联合会在此设培训班培训青年干部（安吴青训班）。

18 ［中华民国］望月楼

望月楼位于吴氏庄园后花园内。建于民国初年，是赏月纳凉之处。2层，坐北朝南，五开间，底层三面回廊，墙面上开券门；券门两侧用圆砖拼出倚柱。二层为七架抬梁式屋架，四坡屋顶，施布纹板、筒瓦，并有简单脊饰。二层四周设回廊。

---

❾ 高兰亭（1895～1958年），泾阳县王桥镇陈家沟人，1919年考入北平大学哲学系，毕业后返乡任教于县师范学校，后参加国民革命军，1937年返乡。时泾干中学因经费问题即将停办，高兰亭被各方推为校长，延请名师，勉力办学，使泾干中学成为渭北地区的名校。

望月楼的立面大量采用券窗、弧形挂落、宝瓶形栏杆等西方建筑元素，反映了民国时期西方建筑对关中地区产生的影响。

19 ［清］孟店周宅

孟店周宅位于三原县鲁桥镇孟店村。周氏为当地富绅。泾阳县吴氏庄园的周氏即是此家的女儿。周宅始建于清嘉庆年间，规模宏大，原有十七座院落，现仅存一座，前后四进，坐北朝南，中轴线上的主要建筑有门楼（左右有塾）、二道门、前厅、正厅和后厅。第一进院落和第三进院落（正厅前）横长，第二进院落（前厅前）和后院都是狭长的、典型的关中"窄院"。正厅为2层楼厅，面阔五间，进深三架梁带前廊，通柱做法。

整个周宅用料讲究，施工质量上乘，木雕、砖雕和石雕工艺精湛，门窗格扇、柱础、壁面、墀头等都布满雕饰。格扇裙板部位的木雕为彩色，主题图案有二十四孝故事，"耕、樵、渔、读"主题，以及"长安八景"等等。

20 ［明］三原城隍庙

三原城隍庙位于三原县城渠岸街中段，始建于明洪武八年（1375年），是我国保存至今的规模最为宏大、格局最为完整的城隍庙建筑群之一。

整个建筑群包括前后五进院落，中轴线上的主要建筑由南至北依次为——砖雕照壁，建于明嘉靖二十九年（1550年）；木牌坊，额题"显佑威灵昭应祠"；门楼；木牌坊，额题"陟降在兹"；石牌坊，额题"明灵保障"；戏楼，建于清乾隆二十二年（1757年）；木牌坊，额题"明灵奠佑"，建于明万历十年（1582年），清咸丰年间重修；献殿和正殿，建于明洪武八年（1375年）。献殿和大殿为勾连搭形式。献殿前左右两侧分立钟楼和鼓楼，亦建于明洪武八年；大殿后为明禋亭，建于明嘉靖三十一年（1552年）；最后为寝殿，建于明成化二十二年（1486年）。自建成至清时，曾多次进行维修、重修及增建，形成今日之规模。

About the Sites

01 [Tang] Stone Sculptures, Chong Mausoleum of Tang, Jingyang County, Xianyang

Located on the south slope of the east section of Cuo 'e Mountain, Chong Mausoleum houses the remains of Emperor Dezong of Tang (r. 780-805) and is now a Key Cultural Heritage Site under State Protection.

Chong Mausoleum made use of the natural mountain in forming the tumulus mound. Its burial chamber rests, at an altitude of 955m, to the east of the highest peak of Cuo 'e Mountain while Yeyu River flows to its south. The site covers an approximately 4,200,000m^2 rectangular-ish piece of ground enclosed by a perimeter of walls, of which only rammed-earth stubs remain today. The whole complex features watchtowers on all four corners and spiritual gateways (*shenmen*) through the perimeter wall open to all cardinal directions. Each of these gateways is flanked at the outside by two stone lions and two gate-towers. Into the south gateway leads the 596m long and 71m wide main spiritual path (*shendao*). On both sides of the path stone sculptures are set, including, from south to north, *huabiaos*, winged horses, ostriches, ceremonial horses and human figures.

02 [Tang] Stone Sculptures, Zhen Mausoleum of Tang, Jingyang County, Xianyang

Located on the south slope of Zhong Mountain, Chong Mausoleum houses the burial site of Emperor Xuanzong of Tang (r. 847-859), and is now a Key Cultural Heritage Site under State Protection.

With Zhong Mountain serving as its tumulus mound, the perimeter walls of Zhen Mausoleum were laid accommodating imposingly the natural rises and falls of the mountain. The length of the wall foundations is 1,680m long at the south, 4,080m north, 2,985m east and 4,440m west. The site features at all four corners watchtowers and to all cardinal directions spiritual gateways, outside each of which two stone lions and two gate-towers were erected. Remains of the offering hall (*xiandian*) was found inside the south gateway, into which leads the 505m long and 68m wide main spiritual path. The path is flanked by stone sculptures, including, from south to north, two *huabiaos*, two winged horses, two ostriches, six ceremonial horses and twelve human figures.

03 [Ming-Qing] Confucian Temple of Jingyang, Xianyang

The Confucian Temple of Jingyang stands on Wenmiao Street. Although its original date of erection remains unclear, a *terminus ante quem* was nonetheless confirmed by *On the Repair of Confucian Temple*, a Song Dynasty tablet inscription which offered an account of a repair project in 1090. After the temple was devastated by the 1556 Shaanxi Earthquake, Zhong Dai, the county head, commissioned its restoration, while later dates such as 1576, 1632, 1764 and the reigning term of Emperors Renzong (r. 1796-1820) and Xuanzong (r. 1821-1850) all witnessed repair jobs delivered to different extents. As a Cultural Heritage Site under Provincial Protection, the temple now serves as the county museum, of which surviving structures include a *Ji-men* (Halbert Gate), a *Dacheng-Dian* (Hall of Great Achivement) as well as east and west wings.

04 [Ming] Mahavira Hall of Taikun Temple, Jingyang County, Xianyang

The Mahavira Hall of Taikun Temple is among the few surviving Ming Dynasty officially-styled timber-structured buildings in the region of Guanzhong. The well-preserved building shows today a generally intact structure and has been listed as a Cultural Heritage Site under Provincial Protection.

Located in the county town of Jingyang, Taikun Temple once served as a temporary imperial residence for Emperor Shizu of Former Qin (r. 357-385) and was transformed to a Buddhist temple, with simultaneously acquiring the presently-known name, no earlier than in Sui Dynasty (581-618). Along the course of history it was renamed several times, to Zhongxing Temple during 743-756 and Huiguo Temple during 976-984. Historical records show a restoration project for the Mahavira Hall along with other structures within the temple complex in 1451 and a fire that spared nothing but the hall in 1862.

The Mahavira Hall demonstrates a happy mixture of the typical Ming Dynasty official style and a vivid local tradition of architecture. The hall houses, *inter alia*, a nearly 3m high stone Buddha manufactured in the days of Northern Dynasties (420-589), which is now a Grade 1 State Cultural Artifact, along with several other finely carved stone Buddha tablets of the same period. Also preserved here are Tang Dynasty stone inscriptions and *On the Repair of Huiguo Temple, Jingyang*, a 1451 inscribed stone tablet.

05 [Qing] Opera Stage of Yingxiang Palace, Jingyang County, Xianyang

The stage of Yingxiang Palace is located in the town of Anwu, Jingyang County while abutting Wu Manor to the east. According to *The Country Records of Jingyang* (1547), the palace was originally built in 1164. Although in 1900s, repair and expansion works were commissioned for the site, the Qing Dynasty stage is now the only surviving structure of the original compound. A 1280 inscribed stone tablet, titled *On the Establishment of the Taoist Palace of Yingxiang*, also stands before the stage today.

The stage sits on the upper floor of a two-storied building facing north. The south façade of the building was rendered a carved-brick gatehouse, with the gateway through its ground floor. The stage itself is three-bay wide, featuring exquisitely carved decorations on its *dougongs*, architraves and wall-coping bricks.

06 [Qing] Weijing Academy, Jingyang County, Xianyang

Before the dissolution of Qing Empire, the region of Guanzhong once celebrated four renowned academies, including Guanzhong Academy in Xi' an, Hongdao Academy in Sanyuan as well as Weijing and Chongshi Academies in Jingyang.

Weijing (Classics Appreciating) Academy was the first in Shaanxi to incorporate western teachings is its curriculum. Listed as a Cultural Heritage Site under Provincial Protection, it is now located inside Yaojia Alley Secondary School in the county town of Jingyang.

The Academy was originally established in 1873. In 1883 its teaching repertoire started to include astronomy, geography, classics, anecdotes, physics, chemistry, mathematics and later, current affairs and foreign languages until its suspension in 1902. The whole site used to cover an area of 170m long and 40m wide while only one lecture hall survives today.

07 [Qing] Chongshi Academy, Jingyang County, Xianyang

Chongshi (Lionizing Truth) Academy once ranked among the four renowned academies of Shaanxi in late Qing period. It is now located in Yaojia Alley, County Town of Jingyang as a Cultural Heritage Site under Provincial Protection.

The Academy was in 1897 formed from the Current Affairs Section of Weijing Academy. It went into suspension in 1902, followed by demolitions of some structures. A lecture hall is now the only surviving structure of the academy.

08 [ROC] Primary School of Hanjiapu, Jingyang County, Xianyang

Located in Hanjiapu Village, Town of Anwu, the Primary School of Hanjiapu ranks among the few ROC educational facilities that have survived till today. The only currently remaining structure, which has fallen into desolation, is a rectangular schoolhouse of which the main entrance was installed on the east flank.

09 [Tang] Zen-Master Wukong Pagoda, Zhenxi Temple, Jingyang County, Xianyang

The Pagoda stands today atop the second highest peak of Cuo'e Mountain, Jingyang. It has been listed as a Cultural Heritage Site under Provincial Protection.

Wukong the Zen-master (735 - 812), born Che Fengchao, was a Jingyang local of some Xianbei descent. In 751 he served in the Tang embassy to the western countries and during the journey stayed in Gandhara where he was afterwards converted. His return to Chang'an happened 38 years later, upon which he took residence in Zhangjing Temple of Chang'an until his death in 812. His remains were returned to Cuo'e Mountain, his homeland, to be buried. To commemorate the Zen-master, the establishment of Zhenxi Temple was then commissioned by Emperor Xianzong (r. 805-820) of Tang, as well as the pagoda later by Emperor Xuanzong (r. 847-860) in 860. Present-day sees, however, only the pagoda but no trace of the temple.

The Pagoda is a multi-storied solid-masonry one. The five-storied structure has a hexagonal plan with a side length of 1.8 meters and a remaining height of 13.3 meters. The foundation, ground floor and lower half of the first floor are of stone blocks while the rest grey bricks. Suffering from severe fractures, appreciable tilting and a heavily damaged top part, the pagoda is currently not in an optimistic state of preservation.

10 [Ming] Chongwen Pagoda, Jingyang County, Xianyang

Located in Chongwen Town, Jingyang County on the north bank of Jing River, Chongwen (Literature Advocating) Pagoda is the second highest standing historic pagoda in China and has been listed as a Key Cultural Heritage Site under State Protection.

The pagoda was originally erected during Emperor Shenzong (r. 1573-1620) of Ming's reign, to invoke heavenly bless for the promotion of literary ethos for Jingyang, Sanyuan and Gaoling, hence the name. The structure was once accompanied by a Buddhist temple, which was demolished in 1862, surviving today only the pagoda.

The multi-storied pagoda was constructed with air bricks. It has an octagonal plan with a side length of 9.3m and thirteen stories with a height of 87.2 meters. It features a central shaft, which together with the exterior walls support sets of stairways leading all the way to the top.

11 [Qing] Dayi *Pailou*, Zhangjiatun Village, Jingyang County, Xianyang

Middled on the thoroughfare leading into Zhangjiatun Village, Yunyang Town, Jingyang, Dayi *Pailou* is the earliest built stone *pailou* still standing in Jingyang.

Supported by four posters and featuring double eaves, the *pailou* has a height of 8.78 meters. It now suffers from heavy damages at the top with some members disintegrated. Many engravings of animal figures, such as dragons, cranes, tigers and lions, could be seen on the *pailou*.

12 [Qing] Defensive Building, Lijia Village, Jingyang County, Xianyang

The defensive building is a Cultural Heritage Site under Provincial Protection located in Lijia Village, Town of Anwu, Jingyang. It is a north-facing five-storied tower-like structure with a height of 22 meters. As a defensive structure against bandits, it has arched windows or loopholes set on all five stories. As said to have been originally erected in Tang period (618-907) and restored in late Qing period, the building epitomizes an architectural type of which a very limited number of examples still remain in Shaanxi Province.

13　[Qing] Former Residence of Li Yizhi, Jingyang County, Xianyang

Located in Sheshu Village, Town of Wangqiao, Jingyang, the Former Residence of Li Yizhi has been listed as a Cultural Heritage Site under Provincial Protection. Of the original residence only one courtyard complex opening to the north remains today, which features a gatehouse, two wings, a stage pavilion and the main hall. The site generally conveys a feeling of commodiousness and openness as well as displays finely carved decorations, especially with the gatehouse and the pavilion. The pavilion stands inside the courtyard and is characterized by a double-rafter round-ridge roof. The five-bay wide main hall has a front veranda demonstrating an analogy with South Chinese vernacular dwellings with its truss structure.

14　[ROC] Residence of Zhao Chunxi, Jingyang County, Xianyang

Located in the Town of Kou, Jingyang, the Residence of Zhao remains today largely intact with a courtyard enclosed by a gatehouse, two wings and a main hall. The gatehouse has a longer front projecting eave than its rear one, which is a typical local treatment. The main hall is three bays wide, six bays long and decorated with fine carvings. The layout and building techniques adopted by the residence could be regarded as representative of the vernacular architectural characteristics prevalent in this region during ROC period (1912-1949).

15　[Qing] Former Residence of Jiang Mingjie, Jingyang County, Xianyang

Located in Jianglu Village, Town of Anwu, Jingyang, Jiang's former residence was originally built in the period between 1661 and 1795. It is now deemed as a representative piece of vernacular architecture of Guanzhong. The residence is south facing, of which only the southernmost courtyard, featuring a gatehouse, two wings and a main hall, remains today. The main hall is three bays wide and four bays long. Finely carved decorations are found on timber members of its front but not those of its rear.

16 [ROC] Former Residence of Gao Lanting, Jingyang County, Xianyang

Gao Lanting (1895-1958) was a Jingyang local. After graduation from the Department of Philosophy, Peking University, he returned to be a teacher for his hometown and later became the principle of Jinggan Secondary School. His former residence, located on Ertiao Street of the county town, is a Jingyang urban dwelling built in the days between Qing and ROC periods. Of the residence the gatehouse, wings and main hall still stand today. The main hall endured later transformation thus deviating now from its original look.

17 [Qing] Wu Manor (the former Youth Training Facility of Anwu), Jingyang County, Xianyang

Located in Anwupu Village, Anwu Town, Wu Manor has been listed as a Key Cultural Heritage Site under State Protection.

The manor was originally established by Wu Weiwen, a salt merchant, at the end of Qing period. His son married a woman (1868-1910), whose maiden name was Zhou, from Mengdian Village, Sanyuan County. Zhou was not only an acute businesswoman that with her superb management skills succeeded in ranking the Wu family among the renowned gentries of Guanzhong, but also a dedicated philanthropist that funded various charitable programs including disaster relief, Confucian temple construction and publishing.

The south-facing manor is 70m long and 30m wide, of which three sets of courtyard complexes remain today. From the entrance inward, an eight-bay entrance hall as well as a passage hall, a main hall and a rear hall, all five bays wide, are aligned successively on the central axis. All of these halls feature longer front projecting eaves than rear ones. This was a widely adopted treatment for vernacular dwellings in Shaanxi Province, which could maximize the natural lighting possibly accessible to the indoor area. Column bases and walls here feature quite finely manufactured masonry carvings as decorations. Such carvings on the gable of the west wing at the rearmost courtyard is exceptionally exquisite, which represents the first-class examples of this kind of works in Guanzhong.

18 [ROC] Wangyue Building, Jingyang County, Xianyang

Located in the rear garden of Wu Manor, Wangyue Building was erected in the early days of ROC period and used for enjoying the coolness and the fine view of the moon.

The south-facing building is two stories high and five bays wide. Its ground floor is surrounded by a veranda on three sides and features arched gateways through the walls. The upper floor is covered by a post-and-lintel supported hip roof and also has veranda on all four sides. Western architectural elements, such as arched windows, curved hanging traceries and vase-shaped balusters, could be seen as widely used for the façade of the tower. This demonstrated the influence western architecture shed on its Guanzhong counterpart in ROC period.

19 [Qing] Zhou Residence of Mengdian, Sanyuan County, Xianyang

Zhou Residence is located in Mengdian Village, Luqiao Town, Sanyuan. The Zhou family, from which the mistress of Wu Manor originated, was among the local gentries at the time. The magnificent mansion was originally built, during the reign of Emperor Renzong of Qing (r. 1796-1820), to feature seventeen courtyard compounds, while only one of these remain today. The surviving compound features four sets of courtyard complexes with the main entrance facing south. Along the central axis of the compound lies a main gateway, a secondary gateway, an entrance hall, a main hall and a rear hall. The residence is a typical Guanzhong vernacular dwelling showing the happy combination of finest locally available building materials and techniques. The decorative carvings of the site, be they on timber, bricks or stones, were all exquisitely delivered.

20 [Ming] Cheng-Huang Temple of Sanyuan, Xianyang

Originally a 1375 erection, the Cheng-Huang Temple of Sanyuan is located in the county town. It is most highly rated for its largest acreage and most intact layout among all the Cheng-Huang temple compounds throughout the country standing today.

The whole compound comprises of the succession of five sets of courtyard complexes. Sitting on the central axis, major structures include, from south to north, a carved-brick screen wall (1550), a timber-structured *pailou*, a gatehouse, another timber-structured *pailou*, a stone *pailou*, a stage tower (1757), a third timber-structured *pailou* (1582), an offering hall, the main hall (1375), Mingyin Pavilion (1552) and the resting chamber (1486). The offering hall is flanked by a bell-tower and a drum-tower at its front. Since its original establishment till today, the compound has been subject to multiple repairs, renovations and expansions.

参考文献
References

1. [宋] 宋敏求. 长安志 [M]. （中国方志丛书·华北地方·二九零号）. 据1931年民国铅印本影印. 二十卷. 长安志图三卷. 台北：成文出版社，1970.
2. [唐] 释圆照. 悟空入竺记 [M]. 东京：大正一切经刊行会，1934.
3. [宋] 赞宁. 宋高僧传 [M]. 范祥雍点校. 北京：中华书局，1987.
4. [清] 刘于义. 雍正敕修陕西通志 [M]. 钦定四库全书本. 一百卷.
5. [清] 刘懋官. 宣统重修泾阳县志 [M]. （中国地方志集成·陕西府县志辑·第七册）. 据宣统三年天津华新印刷局铅印本影印. 十卷. 南京：凤凰出版社. 2007.
6. 陕西省地方志编纂委员会. 陕西省志·第二十四卷 [M]. 西安：三秦出版社，1999.
7. 赵立瀛. 陕西古建筑 [M]. 西安：陕西人民出版社，1992.
8. 咸阳市文物局. 咸阳文物古迹大观 [M]. 西安：三秦出版社，2007.
9. 李慧，曹发展注考. 咸阳碑刻 [M]. 西安：三秦出版社，2003.
10. 黄盛璋. 关于悟空禅师塔铭主要问题辨证 [J]. 文博，1992，06:13-25.
11. 喻梦哲. 明代关中木构遗珠——泾阳太壶寺大雄殿 [M]// 文物建筑（第9辑）. 北京：科学出版社，2017：1-10.

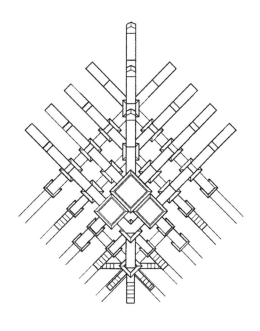

附 录
Appendixes

附录一　本书收录的陕西泾阳、三原地区现存古建筑一览表

序号	名　称	年代	保护级别	地　点
1	唐德宗崇陵石刻	唐	国保（第五批）	泾阳县安吴镇东硷村东北，嵯峨山东段南麓
2	唐宣宗贞陵石刻	唐	国保（第五批）	泾阳县兴隆镇崔黄村北，仲山南麓
3	泾阳文庙	明—清	省保	泾阳县城
4	太壸寺大雄殿	明	省保	泾阳县城
5	迎祥宫戏台	清	无	泾阳县安吴镇安吴堡村
6	味经书院	清	省保	泾阳县城姚家巷中学内
7	崇实书院	清	省保	泾阳县城姚家巷
8	韩家堡小学	中华民国	无	泾阳县安吴镇韩家堡村
9	振锡寺悟空禅师塔	唐	省保	泾阳县安吴镇嵯峨山
10	崇文塔	明	国保（第五批）	泾阳县崇文镇东太平村
11	张家屯村大义坊	清	无	泾阳县云阳镇张家屯村
12	李家村防卫楼	清	省保	泾阳县安吴镇西李家村
13	李仪祉故居	清	省保	泾阳县王桥镇社树村
14	赵春喜民居	中华民国	无	泾阳县口镇
15	蒋明杰民居	清	无	泾阳县安吴镇蒋路村
16	高兰亭故居	中华民国	无	泾阳县城
17	吴氏庄园（含迎祥宫戏台、望月楼）	清	国保（第七批）	泾阳县安吴镇安吴堡村
18	望月楼	中华民国	无	吴氏庄园内
19	孟店周宅	清	省保	三原县鲁桥镇孟店村
20	三原城隍庙	明	国保（第五批）	三原县城

Appendix I: List of Sites

Ref.	Name	TP	Prot. Level	Address
1	Stone Sculptures, Chong Mausoleum of Tang	Tang	State	East section, Cuo'e Mountain, Jingyang
2	Stone Sculptures, Zhen Mausoleum of Tang	Tang	State	South slope, Zhong Mountain, Jingyang
3	Confucian Temple of Jingyang	Ming-Qing	Provincial	County Town of Jingyang
4	Mahavira Hall of Taikun Temple	Ming	Provincial	County Town of Jingyang
5	Opera Stage of Yingxiang Palace	Qing		Anwu Town, Jingyang
6	Weijing Academy	Qing	Provincial	Yaojia Alley, Jingyang
7	Chongshi Academy	Qing	Provincial	Yaojia Alley, Jingyang
8	Primary School of Hanjiapu	ROC		Hanjiapu, Village, Anwu Town, Jingyang
9	Zen-Master Wukong Pagoda	Tang	Provincial	Cuo'e Mountain, Jingyang
10	Chongwen Pagoda	Ming	State	Chongwen Town, Jingyang
11	Dayi Pailou, Zhangjiatun Village	Qing		Zhangjiatun Village, Yunyang Town, Jingyang
12	Defensive Building, Lijia Village	Qing	Provincial	Lijia Village, Anwu Town, Jingyang
13	Former Residence of Li Yizhi	Qing	Provincial	Sheshu Village, Wangqiao Town, Jingyang
14	Residence of Zhao Chunxi	ROC		Kou Town, Jingyang
15	Former Residence of Jiang Mingjie	Qing		Jianglu Village, Anwu Town, Jingyang
16	Former Residence of Gao Lanting	ROC		County Town of Jingyang
17	Wu Manor (including the Stage, Yingxiang Palace and Wangyue Building)	Qing	State	Anwupu Village, Anwu Town, Jingyang
18	Wangyue Building	ROC		Anwupu Village, Anwu Town, Jingyang
19	Zhou Residence of Mengdian	Qing	Provincial	Mengdian Village, Luqiao Town, Sanyuan
20	Cheng-Huang Temple of Sanyuan	Ming	State	County Town of Sanyuan

附录二 图纸目录

1.[唐] 唐德宗崇陵石刻
   1）华表立面图
   2）翼马正面
   3）翼马侧面
   4）文臣正面
   5）文臣侧面
   6）武将正面
   7）武将侧面
   8）石狮大样图

2.[唐] 唐宣宗贞陵石刻
   9）华表立面图
   10）文臣正面
   11）文臣侧面
   12）武将正面
   13）武将侧面
   14）翼马大样图
   15）卧马大样图

3.[明 – 清] 泾阳文庙
   16）总平面图
   17）戟门南立面图
   18）戟门石狮大样图
   19）戟门抱鼓石大样图
   20）戟门北立面图
   21）戟门瓦当大样图
   22）戟门滴水大样图
   23）戟门梁架仰视图
   24）戟门当心间平身科大样图
   25）戟门次间平身科大样图
   26）戟门纵剖面图

27）戟门柱头科大样图

28）戟门横剖面图

29）戟门脊檩下驼峰大样图

30）戟门金檩下驼峰大样图

31）戟门随梁枋大样图

32）戟门额枋大样图

33）大成殿梁架仰视图

34）大成殿南立面图

35）大成殿北立面图

36）大成殿西立面图

37）大成殿次间柱头科大样图

38）大成殿纵剖面图

39）大成殿当心间横剖面图

40）大成殿梢间横剖面图

41）大成殿当心间平身科大样图

42）大成殿次间平身科大样图

43）大成殿梢间平身科大样图

44）大成殿柱础大样图

45）大成殿柱础大样图

4.[明] 太壸寺大雄殿

46）平面图

47）梁架仰视图

48）南立面图

49）东立面图

50）转角铺作大样图

51）3-3 剖面图

52）柱头铺作大样图

53）1-1 剖面图

54）2-2 剖面图

5.[清] 迎祥宫戏台

55）一层平面图

56）北立面当心间平身科大样图

57）南立面图

58）北立面图

59）西立面图

60）砖雕墙头大样图

61）吻兽大样图

6.[清] 味经书院
　　62）平面图
　　63）南立面图

7.[清] 崇实书院
　　64）平面图
　　65）横剖面图

8.[中华民国] 韩家堡小学
　　66）东立面图
　　67）北立面图

9.[唐] 振锡寺悟空禅师塔
　　68）平面图
　　69）南立面图
　　70）二层檐部大样图

10.[明] 崇文塔
　　71）一层平面图
　　72）二至七层平面图
　　73）立面图
　　74）大样图
　　75）八至十三层平面图
　　76）1-1 剖面图
　　77）2-2 剖面图

11.[清] 张家屯村大义坊
　　78）南立面图

79）平面图
80）1-1 剖面图
81）西立面图

12.[清] 李家村防卫楼
82）一层平面图
83）二至五层平面图
84）南立面图
85）西立面图
86）1-1 剖面图

13.[清] 李仪祉故居
87）平面图
88）正厅纵剖面图
89）1-1 剖面图
90）大样图
91）2-2 剖面图
92）入口南立面图

14.[中华民国] 赵春喜民居
93）院落剖面图

15.[清] 蒋明杰民居
94）总平面图
95）东厢房立面图
96）正厅明间横剖面图
97）大样图
98）倒座正立面图

16.[中华民国] 高兰亭故居
99）正立面图
100）平面图
101）当心间横剖面图

17.[清] 吴氏庄园
    102）总平面图
    103）门屋南立面图
    104）正厅南立面图
    105）正厅纵剖面图
    106）正厅明间横剖及厢房立面图
    107）过厅南立面图
    108）过厅纵剖面图
    109）后厅南立面图
    110）后厅西厢房壁门大样图

18.[中华民国] 望月楼
    111）一层平面图
    112）二层平面图
    113）南立面图
    114）1-1 剖面图
    115）东立面图

19.[清] 孟店周宅
    116）总平面图
    117）院落 1-1 剖面图
    118）门屋南立面图
    119）门砧石狮大样图
    120）大门砖雕大样图
    121）二门立面图
    122）院落 2-2 剖面图
    123）怀古月轩南立面图

20.[明] 三原城隍庙
    124）总平面图
    125）院落 1-1 剖面图
    126）照壁北立面图
    127）门楼平面图
    128）门楼南 - 北立面图

129）门楼梁架仰视图
130）门楼横剖面图
131）东西碑廊平面图
132）西碑廊东立面图
133）"明灵奠佑"木牌坊西立面图
134）"明灵奠佑"木牌坊梁架仰视图
135）"明灵奠佑"木牌坊南立面图
136）大殿与献殿平面图
137）大殿梁架仰视图
138）献殿梁架仰视图
139）献殿南立面图
140）献殿东立面图
141）大殿北立面图
142）大殿东立面图
143）寝殿一层平面图
144）寝殿二层平面图
145）寝殿一层梁架仰视图
146）寝殿二层梁架仰视图
147）寝殿院落南立面图
148）寝殿纵剖面图
149）寝殿东立面图
150）寝殿横剖面图

Appendix II: List of Drawings

1. [Tang] Stone Sculptures, Chong Mausoleum of Tang
    1) Elevation, a *huabiao*
    2) Front elevation, a winged horse
    3) Side elevation, a winged horse
    4) Front elevation, a civil servant
    5) Side elevation, a civil servant
    6) Front elevation, a military officer
    7) Side elevation, a military officer
    8) Details, a lion

2. [Tang] Stone Sculptures, Zhen Mausoleum of Tang
    9) Elevation, a *huabiao*
    10) Front elevation, a civil servant
    11) Side elevation, a civil servant
    12) Front elevation, a military officer
    13) Side elevation, a military officer
    14) Details, a winged horse
    15) Details, a crouching horse

3. [Ming - Qing] Confucian Temple of Jingyang
    16) Site plan
    17) South elevation, *Ji-men*
    18) Details, a stone lion at *Ji-men*
    19) Details, a drum-stone of *Ji-men*
    20) North elevation, *Ji-men*
    21) Details, eave tiles of *Ji-men*
    22) Details, a flashing tile of *Ji-men*
    23) Reflected plan, the frame structure of *Ji-men*
    24) Details, an intermediate *dougong* set at the central bay, *Ji-men*
    25) Details, an intermediate *dougong* set at a second-to-the-central bay, *Ji-men*
    26) Longitudinal section, *Ji-men*

27) Details, an on-column *dougong* set, *Ji-men*

28) Transverse section, *Ji-men*

29) Details, a *tuofeng* (camel-hump-shaped support) beneath the ridge purlin, *Ji-men*

30) Details, a *tuofeng* beneath an intermediate purlin, *Ji-men*

31) Details, a tie-beam along the lower edge of a crescent beam, *Ji-men*

32) Details, an architrave, *Ji-men*

33) Reflected plan, the frame structure of *Dacheng-Dian*

34) South elevation, *Dacheng-Dian*

35) North elevation, *Dacheng-Dian*

36) West elevation, *Dacheng-Dian*

37) Details, an on-column *dougong* set at a secondary bay, *Dacheng-Dian*

38) Longitudinal section, *Dacheng-Dian*

39) Transverse section, at the central bay, *Dacheng-Dian*

40) Transverse section, at a lateral bay, *Dacheng-Dian*

41) Details, an intermediate *dougong* set at the central bay, *Dacheng-Dian*

42) Details, an intermediate *dougong* set at a second-to-the-central bay, *Dacheng-Dian*

43) Details, an intermediate *dougong* set at a lateral bay, *Dacheng-Dian*

44) Details, column bases, *Dacheng-Dian* (Part 1)

45) Details, column bases, *Dacheng-Dian* (Part 2)

4. [Ming] Mahavira Hall of Taikun Temple

   46) Plan

   47) Reflected plan, the frame structure

   48) South elevation

   49) East elevation

   50) Details, an on-corner *dougong* set

   51) Section 3-3

   52) Details, an on-column *dougong* set

   53) Section 1-1

   54) Section 2-2

5. [Qing] Opera Stage of Yingxiang Palace

   55) 1F plan

   56) An intermediate *dougong* set at the central bay, north façade

57) South elevation

58) North elevation

59) West elevation

60) Details, the carved-brick wall header

61) Details, a *wenshou* (on-roof beast-shaped ornament)

6. [Qing] Weijing Academy

   62) Plan

   63) South Elevation

7. [Qing] Chongshi Academy

   64) Plan

   65) Transverse section

8. [ROC] Primary School of Hanjiapu

   66) East elevation

   67) North elevation

9. [Tang] Zen-Master Wukong Pagoda

   68) Plan

   69) South elavation

   70) Details, 2F eaves

10. [Ming] Chongwen Pagoda

   71) 1F Plan

   72) 2-7F Plans

   73) Elevation

   74) Details

   75) 8-13F Plans

   76) Section 1-1

   77) Section 2-2

11. [Qing] Dayi *Pailou*, Zhangjiatun Village

   78) South elevation

79) Plan
80) Section 1-1
81) West elevation

12. [Qing] Defensive Building, Lijia Village
    82) 1F Plan
    83) 2-5F Plans
    84) South elevation
    85) West elevation
    86) Section 1-1

13. [Qing] Former Residence of Li Yizhi
    87) Plan
    88) Longitudinal section, the main hall
    89) Section 1-1
    90) Details
    91) Section 2-2
    92) South elevation, entrance

14. [ROC] Residence of Zhao Chunxi
    93) Cross section, the courtyard complex

15. [Qing] Former Residence of Jiang Mingjie
    94) Site plan
    95) Elevation, the east wing
    96) Transverse section, at the central bay, the main hall
    97) Details
    98) Elevation, the *daozuo* (the house facing the main hall in a *siheyuan*)

16. [ROC] Former Residence of Gao Lanting
    99) Elevation
    100) Plan
    101) Transverse section, at the central bay

17. [Qing] Wu Manor

   102) Site Plan

   103) South elevation, the gatehouse

   104) South elevation, the main hall

   105) Longitudinal section, the main hall

   106) Transverse section, at the central bay of the main hall, plus the elevation of a wing

   107) South elevation, the passage hall

   108) Longitudinal section, the passage hall

   109) South elevation, the rear hall

   110) Details, a *bimen* (an ornate gateway through the walled lateral end of the veranda)

18. [ROC] Wangyue Building

   111) 1F Plan

   112) 2F Plan

   113) South elevation

   114) Section 1-1

   115) East elevation

19. [Qing] Zhou Residence of Mengdian

   116) Site plan

   117) Section 1-1, the courtyard complex

   118) South elevation, the gatehouse

   119) Details, stone lions as bearing-stones, the entrance

   120) Details, brick carvings, the entrance

   121) Elevation, the secondary gateway

   122) Section 2-2, the courtyard complex

   123) South elevation, Huaiguyue Pavilion

20. [Ming] Cheng-Huang Temple of Sanyuan

   124) Site plan

   125) Section 1-1, the courtyard complex

   126) North elevation, the screen wall

   127) Plan, the gatehouse

   128) South / north elevations, the gatehouse

129) Reflected plan, the frame structure of the gatehouse
130) Transverse section, the gatehouse
131) Plan, the east and west stele galleries
132) East elevation, the west stele gallery
133) West elevation, the '*Ming-Ling-Dian-You* (Blessed by Brilliant Spirits)' timber-structured *pailou*
134) Reflected plan, the frame structure of the '*Ming-Ling-Dian-You*' *pailou*
135) South elevation, the '*Ming-Ling-Dian-You*' *pailou*
136) Plans, the main hall and the offering hall
137) Reflected plan, the frame structure of the main hall
138) Reflected plan, the frame structure of the offering hall
139) South elevation, the offering hall
140) East elevation, the offering hall
141) North elevation, the main hall
142) East elevation, the main hall
143) 1F Plan, the resting chamber
144) 2F Plan, the resting chamber
145) Reflected plan, the frame structure, 1F, the resting chamber
146) Reflected plan, the frame structure, 2F, the resting chamber
147) South elevation, the resting chamber courtyard
148) Longitudinal section, the resting chamber
149) East elevation, the resting chamber
150) Transverse section, the resting chamber

后　记

　　1995年，我开始从事中国建筑史的教学工作，1998年恢复了停开已久的"古建筑测绘"课程。那一年的秋天，我和建筑学1995级的同学们远赴陕北佳县，测绘了坐落于白云山上的道教建筑群白云观，这是我测绘教学生涯的开始，至今整整20年了。

　　起初相当长的一段时间里，承乏测绘指导老师者仅我一人，其后渐有越来越多的青年教师参与进来，如今已经形成了一个稳定的测绘教学与科研团队。在完成测绘教学任务之余，我们有计划地对陕西地区的大量建筑遗产进行了测绘记录与后续研究，并间及其他省区，如甘肃、山西、西藏、辽宁等地的多处古建筑，类型涵盖祠庙、帝陵、佛寺、道观、清真寺、书院、会馆、佛塔、城垣、民居与古村落以及历史街区，还有近代的教堂建筑等等。我校开办"风景园林学"本科专业后，结合我们的中国园林史研究方向，为该专业新开设了"古典园林测绘"课程，2011年至今已测绘了江南地区的九处古典园林。

　　在测绘成果长期大量积累的基础上，我们进行了地域建筑史、技术史、园林史与遗产保护等方向的持续研究，撰写、出版了《古建筑测绘学》、《凌苍莽·瞰紫微——陕西古塔实录》[三卷本，分别为隋唐时期、宋（金）元时期和明清时期]、《中国古建筑测绘大系·陕西祠庙》和《中国古建筑测绘大系·江南古典园林》（"十二五"国家重点图书出版规划项目）以及《苏州艺圃》等专著，并有多篇研究论文发表，亦获准多项国家自然科学基金、陕西省社会科学基金及其他省部级研究项目。

　　这20年间的测绘教学与研究工作，初期着重选取的是那些较具重要性与代表性的建筑遗产，此为志在廓清陕西地区建筑遗产整体认知版图的基础性工作。而最近10年，我们则更着力于对这一版图进行丰富和细化，消除其中的盲点，使之越发清晰。

　　陕西建筑遗产的类型和数量均十分丰富，在很多地区不仅分布集中，而且价值较高、保存状况普遍较好。鉴于省内建筑遗产的这一现状特点，我们产生了编写陕西境内各地区建筑遗产实录的构想，也想以此为纲系统地整理数量极为可观的历年测绘成果，于是以地区为单位的田野调查和测绘记录就此展开，并结合第三次全国文物普查的新发现，在已有的工作基础上有针对性地查缺补漏，以求尽量补全各地区建筑遗产的测绘成果。虽然受各遗产点外业条件以及管理、使用状况等现实因素所限，毫发无遗的美好设想难以实现，但是只要条件允许，我们都尽可能地进行了测绘记录。当然，这个编写陕西各地区建筑遗产实录系列丛书的计划需有长期的积累和不懈的坚持方可实现。

　　1992年，我的导师赵立瀛教授主编的《陕西古建筑》出版，作为对陕西地区古建筑第一次全面记录研究的成果，该书直至今天仍然是研究陕西地区古建筑的代表性著作。对于前辈学者的成果，我们今日所为，既欲克绍箕裘，亦求发扬光大。为表达我校建筑历史与理论学科的这一学术传承，我们仍以"陕西古建筑"为主题，将丛书定名为《陕西古建筑测绘图辑》，

后 记

此"泾阳·三原"卷即是我们第一阶段已经完成的部分工作成果。

泾阳和三原两县毗邻，隶属咸阳地区，北距西安40余公里。自2014年春季开始至年底，我们对泾阳、三原两地集中进行了建筑遗产测绘与田野调查，在当地政府的支持下泾阳县境内的测绘工作做得较为全面，三原县则主要测绘了最负盛名的两处建筑群——三原城隍庙和孟店周宅。

本书在测绘成果的整理和表达、图纸的选取与版面设计等方面，进行了一些探索和尝试。此举旨在避免测绘图的面目过于严肃和专业，乃至于将非专业读者拒之门外。全书内容以图为主体，并为保证读图的连续性，将说明文字集中于附录部分，文字本身也力求言简意赅。选取图纸时，以平面图、正立面图、横剖面图和建筑组群纵剖面图等直观的、观赏性较强的且又能较好地呈现建筑之特征的图为主，适量选取了记录、表达测绘对象最具价值的或最富特色的细部大样图以及现状照片。总体而言，我们的目标在于使建筑测绘图在保证专业性的前提下更为好看和易读。当然了，是否达到了这一目标有待读者诸君的检验。

建筑学院院长刘加平院士和我的导师赵立瀛先生慨然应允，为本书作序，对此至为感谢！

感谢本书的责任编辑戚琳琳女士和率琦先生，他们的专业、高效和辛劳付出使得本书在短短数月内即付梓面世。

是为记。

林源

2017年10月30日
于西安

Postscript

I became an architectural history teacher in 1995 and reinstated Historic Building Metric Survey, a long suspended course for student architects three years later. In the autumn of 1998, undergraduate architects from the Class of 2000 and I went to Jia County and put out recording tapes on Taoist buildings of the Baiyun Mountain. It was that fieldtrip that initiated me into the role of a metric survey teacher, since which twenty years have passed.

In a rather long period of time since then, I served as the only teacher for the course. But more and more new bloods joined me later in the ranks and we eventually had a regular team of metric survey instructors and researchers ready for work. While accomplishing our teaching tasks, we methodically made continuous efforts on surveying and follow-up studying on the vast array of standing historic buildings in the province, along with many others elsewhere, e.g. Gansu, Shanxi, Tibet and Liaoning. These buildings and sites included Buddhist and Taoist temples, imperial mausoleums, mosques, churches, ancient academies, guild houses, pagodas, ramparts, dwellings as well as villages and historic urban areas. Later the university started the undergraduate program of Landscape Gardening, and laid down the course Classical Garden Metric Survey for the pathway of Chinese Gardening History offered by this program. For this course we have also worked on nine classical gardens in Jiangnan since 2011.

With the multitudinous volumes of materials we have heretofore gathered through the survey jobs, we conducted years of studies on regional architectural history, technical history, gardening history and cultural heritage conservation. The published results of these studies include *Metric Survey for Historic Buildings* (2003), *Over the Land the Approaching the Stars: Shaanxi Historic Pagoda Surveyed* (2016) (three volumes: Vol.1. Sui & Tang, Vol 2. Song, Jin & Yuan and Vol 3. Ming & Qing), *A Compendium of Ancient Chinese Architecture Measured and Drawn: Shaanxi Temples* (2017), *A Compendium of Ancient Chinese Architecture Measured and Drawn: Classical Gardens in Jiangnan* (2018) (a Twelve-Five Year Plan Key Publication), and *The Garden of Cultivation in Suzhou* (2017) etc., as well as many journal papers. We were also granted research funds by National Natural Science Foundation of China (NSFC), Shaanxi Provincial Foundation for Social Sciences Studies, and other provincial and ministerial funds.

These twenty years of work started by choosing almost exclusively subjects of exceptional value and repute. This was deemed as necessary groundbreaking work for grasping the key features in our cognitive map of architectural heritage in the province. The recent ten years, however, saw us especially focus on enhancing and clarifying that map, i.e. filling gaps and erasing blind spots.

The Province of Shaanxi is blessed with a distinctively rich inventory of many architectural heritage sites of many types. Some regions even boast its concentration of high-value and generally well-preserved samples. It was such a realization that actuated us to bring on our agenda compiling a series of first-hand gleaned records on these heritages, which also offered us a chance to systemize our survey results accumulated through the years. A regional based project of architectural heritage surveying was thus launched. Availed of the new findings of the 3rd National Reconnaissance of Cultural Heritage, we purposefully chose a gap-filling strategy, which means working to complete our already acquired records region by region. Although we intended to leave no stone unturned, conditions, including diverse physical, managerial and practical situations of every given site, impeded our fulfilling thoroughly that intention. Nonetheless we completed our tasks whenever and wherever we were able to. It is beyond doubt that the completion of this project, that is, creating a series of measured and drawn records for architectural heritage of every region in the province, will entail accumulation and dedication over a long period of time.

In 1992, *Shaanxi Historic Buildings* (ISBN: 97872240149451), by my mentor Prof. Zhao Liying, was published. As the first comprehensive recording and study on the historic buildings in Shaanxi, the book has remained till today the first choice for reference on the subject. The intellectual legacy of our predecessors we aim now not only to inherit but also enhance. To pay homage to the academic lineage of architectural historians in the university we chose to keep the keywords of Shaanxi Historic Buildings in our series title, which has been finally determined to be *Shaanxi Historic Buildings Measured and Drawn. Of the series Volume Jingyang & Sanyuan is* now lying before you, which presents the part of work we have done for the first stage of the project.

Jingyang and Sanyuan, two contiguous counties under Xianyang, are 40 kilometers north of Xi'an. Throughout 2014 we carried out intensive metric surveying work for historic buildings in the two places. Thanks to the collaboration and assistance offered by the local government of Jingyang, our work turned out rather exhaustive there. While in Sanyuan we mainly focused on the two most distinguished sites: Cheng-Huang Temple and Zhou Residence of Mengdian.

About presenting this book, such as content organizing, drawing selection and layout design, we also venture to propose some unconventional ideas. We try not to pose these drawings as too 'serious' or professional lest they shut out potential readers outside our line of business. To guarantee our readers' uninterrupted enjoyment of the drawings, which constitutes the bulk of the volume, we grouped introductory texts for each site together at the last section of the book.

These texts themselves were also rendered as concise as possible for that purpose. For drawings we prioritized intuitively and aesthetically enjoyable ones that could highlight each site's characteristics, which mainly included plans, elevations and transverse sections for individual buildings and longitudinal sections for groups. We also added some details and photographs instrumental in showing each site's most distinctive and highly-valued features. In a nutshell, our goal is to offer architectural drawings in a well-presentable and easily-readable form without compromising their capacity of meeting professional standards. Whether this volume has achieved that goal, we leave to the judicious judgement of our readers.

Prof. Liu Jiaping, a CAE academician and the Dean of School of Architecture, XAUAT, and Prof. Zhao Liying, my mentor, both deigned to write a foreword for this series. We are greatly honored by and particularly grateful to their endorsement.

Special thanks would go to Ms. Qi Linlin and Mr. Shuai Qi from China Architecture & Building Press (CABP), the executive editors of this volume. The publishing process would not have merely taken a few months without their expertise, efficiency and efforts.

Hence we are honored to present you this volume.
Sincerely yours,

Ocrober 30th, 2017, in Xi'an

(English translation by Lin Xi)